Collins

The Shanghai Maths Project

For the English National Curriculum

Practice Book 3A

Series Editor: Professor Lianghuo Fan

UK Curriculum Consultant: Paul Broadbent

William Collins' dream of knowledge for all began with the publication of his first book in 1819.

A self-educated mill worker, he not only enriched millions of lives, but also founded a flourishing publishing house. Today, staying true to this spirit, Collins books are packed with inspiration, innovation and practical expertise. They place you at the centre of a world of possibility and give you exactly what you need to explore it.

Collins. Freedom to teach.

Published by Collins
An imprint of HarperCollins*Publishers*
The News Building
1 London Bridge Street
London
SE1 9GF

Browse the complete Collins catalogue at
www.collins.co.uk

HarperCollins*Publishers*
Macken House, 39/40 Mayor Street Upper,
Dublin 1, D01 C9W8, Ireland

© HarperCollins*Publishers* Limited 2017

© Professor Lianghuo Fan 2017

© East China Normal University Press Ltd. 2017

10 9 8 7 6

ISBN 978-0-00-822611-4

Translated by Professor Lianghuo Fan, Adapted by Professor Lianghuo Fan.

British Library Cataloguing in Publication Data

A catalogue record for this publication is available from the British Library.

Series Editor: Professor Lianghuo Fan
UK Curriculum Consultant: Paul Broadbent
Publishing Manager: Fiona McGlade
In-house Editor: Nina Smith
In-house Editorial Assistant: August Stevens
Project Manager: Emily Hooton
Copy Editors: Catherine Dakin and Karen Williams
Proofreader: Jo Kemp
Cover design: Kevin Robbins and East China Normal University Press Ltd.
Cover artwork: Daniela Geremia
Internal design: 2Hoots Publishing Services Ltd
Typesetting: 2Hoots Publishing Services Ltd
Illustrations: QBS
Production: Rachel Weaver
Printed and bound in the UK using 100% Renewable Electricity at CPI Group (UK) Ltd

The Shanghai Maths Project (for the English National Curriculum) is a collaborative effort between HarperCollins, East China Normal University Press Ltd. and Professor Lianghuo Fan and his team. Based on the latest edition of the award-winning series of learning resource books, *One Lesson, One Exercise*, by East China Normal University Press Ltd. in Chinese, the series of Practice Books is published by HarperCollins after adaptation following the English National Curriculum.

Practice Book Year 3A has been translated and developed by Professor Lianghuo Fan with the assistance of Ellen Chen, Ming Ni, Huiping Xu and Dr Lionel Pereira-Mendoza, with Paul Broadbent as UK Curriculum Consultant.

This book is produced from independently certified FSC™ paper to ensure responsible forest management.
For more information visit: www.harpercollins.co.uk/green

Contents

Chapter 1 Revising and improving

1.1 Revision for addition and subtraction of 2-digit numbers

 Learning objective Add and subtract 2-digit numbers

 Basic questions

1 Calculate mentally.

(a) 63 + 9 = ☐   (b) 75 + 24 = ☐   (c) 84 − 16 = ☐

(d) 70 − 35 = ☐   (e) 34 + 16 = ☐   (f) 48 + 25 = ☐

(g) 63 − 36 = ☐   (h) 93 − 53 = ☐   (i) 38 + 25 = ☐

(j) 27 + 44 = ☐   (k) 74 − 58 = ☐   (l) 60 − 42 = ☐

(m) 36 + 18 = ☐   (n) 67 − 27 = ☐   (o) 50 − 43 = ☐

(p) 90 + 10 = ☐   (q) 82 − 45 = ☐   (r) 36 + 55 = ☐

2 Add three numbers.

(a) 21 + 23 + 50 = ☐   (b) 36 + 44 + 13 = ☐

(c) 13 + 30 + 37 = ☐   (d) 23 + 20 + 40 = ☐

(e) 67 + 25 + 12 = ☐   (f) 42 + 26 + 24 = ☐

(g) 60 + 18 + 11 = ☐   (h) 21 + 44 + 12 = ☐

(i) 35 + 16 + 40 = ☐   (j) 34 + 35 + 14 = ☐

3 Subtract three numbers.

(a) 72 − 33 − 12 =

(b) 54 − 27 − 18 =

(c) 67 − 21 − 22 =

(d) 68 − 45 − 18 =

(e) 65 − 56 − 3 =

(f) 86 − 35 − 24 =

(g) 88 − 12 − 51 =

(h) 79 − 19 − 26 =

(i) 77 − 56 − 6 =

(j) 89 − 42 − 13 =

4 Complete these mixed addition and subtraction calculations.

(a) 87 + 10 − 27 =

(b) 57 + 31 − 27 =

(c) 54 − 52 + 17 =

(d) 96 − 63 + 41 =

(e) 74 − 47 + 15 =

(f) 88 − 46 + 55 =

(g) 54 − 33 + 21 =

(h) 36 − 18 + 9 =

(i) 77 + 18 − 34 =

(j) 64 + 23 − 19 =

5 Fill in the boxes.

(a) ☐ + 41 = 56

(b) 82 − ☐ = 55

(c) ☐ − 36 = 39

(d) 36 + ☐ = 71

(e) 80 − ☐ = 29

(f) 34 = ☐ − 34

6 Application problems.

£18   £15   £46   £25

(a) Zoe bought a 🏀 and a ⚽. How much did they cost her in total?

(b) David paid for a ⌚ with a 50 pound note. How much change did he get?

(c) Mia bought a 🏀 and a ⌚.

How much cheaper is the 🏀 than the ⌚?

Challenge and extension question

7 Write + or − in each circle to make each equation true.

(a) 1 ◯ 2 ◯ 3 ◯ 5 ◯ 4 = 1

(b) 1 ◯ 2 ◯ 3 ◯ 4 ◯ 5 ◯ 6 = 1

1.2 Addition and subtraction (1)

Learning objective Add and subtract 2-digit numbers

 Basic questions

1 Calculate mentally.

(a) 35 + 8 = ☐

(b) 54 + 5 = ☐

(c) 91 − 9 = ☐

(d) 63 − 8 = ☐

(e) 35 + 65 = ☐

(f) 42 + 9 = ☐

(g) 53 − 6 = ☐

(h) 74 − 60 − 3 = ☐

(i) 40 + 50 − 5 = ☐

(j) ☐ − 35 = 20

(k) ☐ + 81 = 81

(l) ☐ − 7 + 5 = 44

2 Use the column method to calculate.

(a) 36 + 48 = ☐

(b) 82 − 25 = ☐

(c) 70 − 32 + 49 = ☐

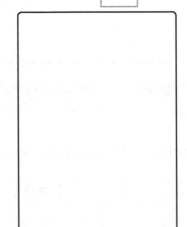

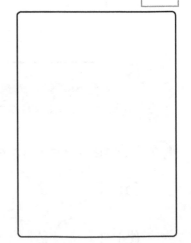

4

3 Write each number sentence and calculate.

(a) How many pieces of fruit
are there?

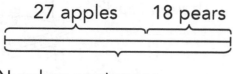

27 apples   18 pears

Number sentence:

_____

(b) There are 27 apples. How
many pears are there?

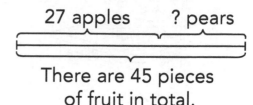

27 apples   ? pears

There are 45 pieces
of fruit in total.

Number sentence:

_____

(c) How many more girls are
there?

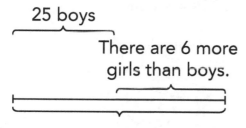

25 boys

There are 6 more
girls than boys.

Number sentence:

_____

(d) How many more boys are
there than girls?

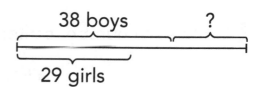

38 boys   ?

29 girls

Number sentence:

_____

4 Write each number sentence and calculate.

(a) Grandma bought 56 eggs. After baking cakes, she had 19 eggs
left. How many eggs did Grandma use?

Number sentence: _____

Answer: _____

(b)

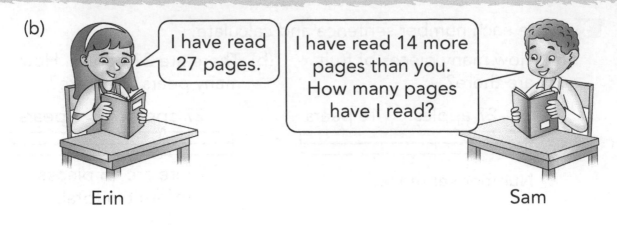

Erin: I have read 27 pages.

Sam: I have read 14 more pages than you. How many pages have I read?

Number sentence: _____

Answer: _____

Challenge and extension question

5 Write each number sentence and calculate.

£47   £24   £38   £76

(a) How much cheaper is the toy plane than the toy car?

Number sentence: _____

(b) If you had £100 to spend, and you wanted to spend as much of it as possible, which toys could you buy? How much change would you get?

_____

_____

_____

1.3 Addition and subtraction (2)

 Learning objective Add and subtract 2-digit numbers

 Basic questions

1 Calculate mentally.

(a) $47 - 25 = \boxed{}$

(b) $76 + 18 = \boxed{}$

(c) $32 + 49 = \boxed{}$

(d) $18 + 49 = \boxed{}$

(e) $9 + 81 = \boxed{}$

(f) $15 + 47 = \boxed{}$

(g) $82 - 16 = \boxed{}$

(h) $38 - 20 = \boxed{}$

(i) $65 - 19 = \boxed{}$

(j) $51 - 7 = \boxed{}$

(k) $77 - 14 = \boxed{}$

(l) $36 - 18 = \boxed{}$

2 Fill in the missing numbers.

(a) $34 + 47 = \boxed{}$

(b) $\boxed{} + 47 = 71$

(c) $\boxed{} + 47 = 61$

(d) $34 + \boxed{} = 51$

(e) $93 - 29 = \boxed{}$

(f) $93 - \boxed{} = 74$

(g) $93 - \boxed{} = 54$

(h) $93 - \boxed{} = 34$

(i) $\boxed{} - 25 = 39$

(j) $\boxed{} - 35 = 39$

(k) $\boxed{} - 45 = 39$

(l) $80 - \boxed{} = 39$

3 Write each number sentence and calculate.

(a)

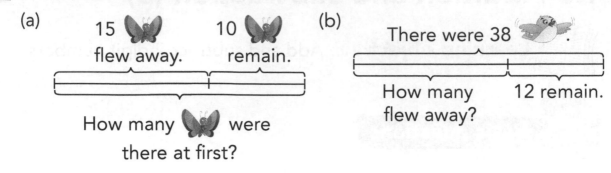

15 flew away.   10 remain.

How many were there at first?

Number sentence:

_____

(b)

There were 38.

How many flew away?   12 remain.

Number sentence:

_____

(c)

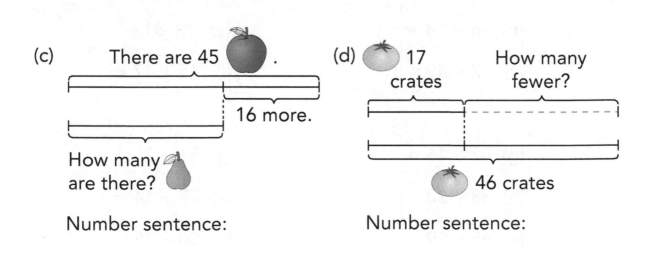

There are 45.

16 more.

How many are there?

Number sentence:

_____

(d)

17 crates   How many fewer?

46 crates

Number sentence:

_____

4 Application problems.

£24   £32   £68   £48

(a) Tom bought two toys for exactly £100. Which toys did he buy?

Number sentence: _____

(b) Asif bought a toy car and a skateboard.

How much did he spend?

Number sentence: _____

(c) Anya had £54 and bought an art set. How much money did she have left?

Number sentence: _____

Challenge and extension question

5 Replace the letters with non-zero numbers to make each calculation correct.

$$
\begin{array}{r}
M\ M \\
+\quad\ M \\
\hline
8\ \ 4
\end{array}
\qquad\qquad
\begin{array}{r}
A\ T \\
-\ T\ A \\
\hline
7\ \ 2
\end{array}
$$

M = ☐   A = ☐   T = ☐

1.4 Calculating smartly

Learning objective Use strategies to add and subtract 2-digit numbers

Basic questions

1 Fill in the boxes.

(a) 38 + [ ] = 40   (b) 60 + [ ] = 70   (c) [ ] + 54 = 60

(d) [ ] + 71 = 80   (e) 45 + [ ] = 50   (f) 26 + [ ] = 30

2 Calculate smartly and fill in the boxes.

(a)
48 + 17 = [ ]
↓ ↓ ↑
49 + 16
↓ ↓
50 + 15 = [ ]

(b)
55 + 27 = [ ]
↓ ↓ ↑
54 + 28
↓ ↓
53 + 29
↓ ↓
52 + 30 = [ ]

(c)
72 − 33 = [ ]
↓ ↓ ↑
71 − 32
↓ ↓
70 − 31
↓ ↓
69 − 30 = [ ]

(d)
64 − 17 = [ ]
↓ ↓ ↑
65 − 18
↓ ↓
66 − 19
↓ ↓
67 − 20 = [ ]

(e)
49 + 16 = [ ]
↓+1 ↓−1
50 + 15 = [ ]

(f)
75 − 36 = [ ]
↓+4 ↓+4
79 − 40 = [ ]

(g)

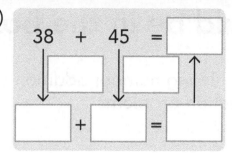

38 + 45 =

(h)

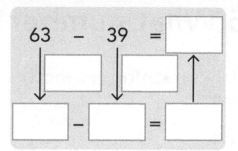

63 – 39 =

3 Fill in the boxes to complete each calculation.

(a) 38 + 14 = 40 + ☐ = ☐   (b) 56 – 27 = ☐ – 30 = ☐

(c) 17 + 49 = ☐ + ☐ = ☐   (d) 93 – 47 = ☐ – ☐ = ☐

(e) ☐ + 28 = 44 + ☐ = 74   (f) 73 – ☐ = ☐ – 30 = ☐

(g) 24 + 69 = ☐ + ☐ = ☐ + ☐ = ☐ + ☐ = ☐ + ☐

(h) 72 – 34 = ☐ – ☐ = ☐ – ☐ = ☐ – ☐ = ☐ – ☐

4 Draw lines to match the calculations that have the same answer.

55 + 28 ●   ● 60 – 13

13 + 48 ●   ● 53 + 30

62 – 15 ●   ● 78 – 40

74 – 36 ●   ● 10 + 51

Challenge and extension question

5 Ethan was adding two numbers. He mistook the digit 7 for 5 in the ones place of one of the addends and the digit 4 for 6 in its tens place. This gave him an answer of 92.

The correct answer should be ☐ .

1.5 What number should be in the box?

Learning objective Solve missing number addition and subtraction problems

Basic questions

1 Fill in the boxes.

(a) $52 +$ ☐ $= 81$ (b) $47 -$ ☐ $= 9$ (c) ☐ $- 25 = 39$

(d) $81 - 52 =$ ☐   (e) $47 - 9 =$ ☐   (f) $39 + 25 =$ ☐

(g) $33 +$ ☐ $= 67$ (h) $56 -$ ☐ $= 28$ (i) ☐ $- 24 = 44$

(j) ☐ $+ 18 = 18$ (k) ☐ $- 37 = 63$ (l) $67 -$ ☐ $= 28$

(m) $64 -$ ☐ $= 34$ (n) ☐ $- 73 = 19$ (o) $48 +$ ☐ $= 82$

2 Look at the diagrams and fill in the boxes.

(a)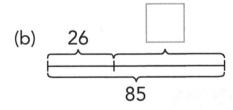

☐ $+$ ☐ $=$ ☐   ☐ $+$ ☐ $=$ ☐

☐ $-$ ☐ $=$ ☐   ☐ $-$ ☐ $=$ ☐

(b)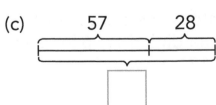

$26 +$ ☐ $= 85$
↑

$85 - 26 =$ ☐

(c)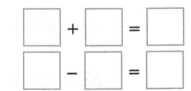

☐ $- 57 = 28$
↑

$57 + 28 =$ ☐

(d)

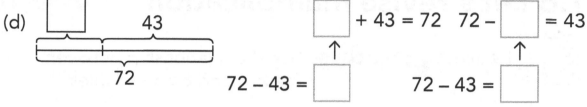

$\square + 43 = 72$ $72 - \square = 43$

$72 - 43 = \square$ $72 - 43 = \square$

3 Write each number sentence and calculate.

(a) There were 72  in the pond at first.

34 were left.   How many swam away?

Number sentence: _____

(b) There are 15 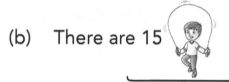   There are 36

How many children are there altogether?

Number sentence: _____

(c) There were 12 in a tree.   How many more joined them?

There are now 21 birds.

Number sentence: _____

 **Challenge and extension question**

4 Look at each number sentence. Then choose > or < to write in the circle below.

(a) ■ + 17 = 26 + ●   (b) ■ − 9 = ● − 15   (c) ■ + 8 = ● − 8

■ ◯ ●   ■ ◯ ●   ■ ◯ ●

1.6 Let's revise multiplication

Learning objective Use the relationship between the 2, 4 and 8 times tables

Basic questions

1 Write the multiplication facts.

(a) Fill in the multiplication table.

(b) Circle the multiplication facts that can be used to write only one multiplication sentence and one division sentence.

(c) Using the same coloured pen, colour the multiplication facts with the same products.

1 × 1 = 1				
1 × 2 = 2	2 × 2 = 4			
	2 × 3 = 6			
		3 × 4 = 12	4 × 4 = 16	
				5 × 5 = 25
1 × 8 = 8				
1 × 9 = 9				

(d) Use the multiplication table to find the relationship between multiplications of 2, 4 and 8.

(e) Can you get any other results by looking at the multiplication table?

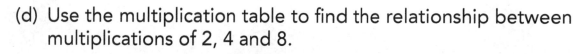

		8 × 8 = 64		
			9 × 10 = 90	10×10=100

2 Draw lines to help each cat find the right house.

8 × 9

10 ÷ 5

2 times 5 is 10

24 ÷ 8

3 × 8

3 times 8 is 24

60 ÷ 10

60 ÷ 6

6 times 10 is 60

6 × 10

72 ÷ 9

8 times 9 is 72

3 Use the multiplication facts to fill in the boxes.

(a) The product is 20.

☐ × ☐ = 20   ☐ × ☐ = 20

☐ × ☐ = 20   ☐ × ☐ = 20

(b) The product is 24.

☐ × ☐ = 24   ☐ × ☐ = 24

☐ × ☐ = 24   ☐ × ☐ = 24

(c) The product is 30.

☐ × ☐ = 30   ☐ × ☐ = 30

☐ × ☐ = 30   ☐ × ☐ = 30

(d) The quotient is 3.

$$\boxed{} \div \boxed{} = 3 \qquad\qquad \boxed{} \div \boxed{} = 3$$

$$\boxed{} \div \boxed{} = 3 \qquad\qquad \boxed{} \div \boxed{} = 3$$

(e) The quotient is 5.

$$\boxed{} \div \boxed{} = 5 \qquad\qquad \boxed{} \div \boxed{} = 5$$

$$\boxed{} \div \boxed{} = 5 \qquad\qquad \boxed{} \div \boxed{} = 5$$

(f) The quotient is 8.

$$\boxed{} \div \boxed{} = 8 \qquad\qquad \boxed{} \div \boxed{} = 8$$

$$\boxed{} \div \boxed{} = 8 \qquad\qquad \boxed{} \div \boxed{} = 8$$

Challenge and extension question

4 Fill in the missing numbers.

(a) $\boxed{} \times 5 = \boxed{} \times 10$

(b) $\boxed{} \times 2 = \boxed{} \times 4 = \boxed{} \times 8$

(c) $\boxed{} \times 3 = \boxed{} \times 6 = \boxed{} \times 9$

1.7 Games of multiplication and division

Learning objective Use multiplication and division facts to solve problems

Basic questions

1 Look at the diagrams and fill in the boxes. Write the number sentence underneath each one.

(a)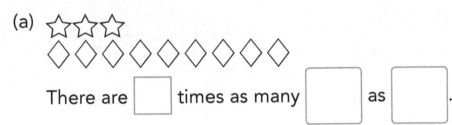

There are ☐ times as many ☐ as ☐ .

Number sentence: _____

(b)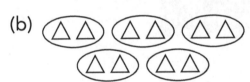

There are ☐ groups of ☐ in ☐ .

Number sentence: _____

(c) How many footballs are there in total? Group the footballs in two different ways and write the number sentences.

Number sentence:

☐ × ☐ + ☐ = ☐

Number sentence:

☐ × ☐ + ☐ = ☐

Number sentence:

☐ × ☐ − ☐ = ☐

Number sentence:

☐ × ☐ − ☐ = ☐

2 Write each number sentence and find the answer.

(a) How many days will the apples last if 6 are eaten each day?

Number sentence: _____

(b) How many days will the apples last if 3 are eaten each day?

Number sentence: _____

(c) How many days will the apples last if 7 are eaten each day?

How many apples will be left over on the last day?

Number sentence: _____

3 Draw the number of shapes as indicated and then fill in the number sentences below.

In the first row,
draw 2 ☆.

In the second row,
draw 4 △.

In the third row, there
are 4 times as many ◯
as ☆ in the first row.

(a) There are [ ] times as many ◯ as △.

[ ] × [ ] = [ ]

(b) 5 times the number of ◯ is [ ].

[ ] × [ ] = [ ]

19

Challenge and extension question

4 All the pupils in a class are assembled on the sports field. They form exactly 5 rows of the same number of pupils. Tom is in the second row. He is in the fourth place from the left. He is also in the fourth place from the right. How many pupils are there in the class?

Chapter 1 test

1 Calculate mentally.

(a) $46 + 19 = \boxed{}$ (b) $54 - 45 = \boxed{}$ (c) $1 \times 10 = \boxed{}$

(d) $10 \div 5 = \boxed{}$ (e) $32 - 7 = \boxed{}$ (f) $67 + 30 = \boxed{}$

(g) $4 \times 12 = \boxed{}$ (h) $36 \div 9 = \boxed{}$ (i) $63 - 42 = \boxed{}$

(j) $72 + 17 = \boxed{}$ (k) $5 \times 8 = \boxed{}$ (l) $35 \div 5 = \boxed{}$

2 Complete the mixed addition and subtraction calculations.

(a) $51 - 42 + 53 = \boxed{}$ (b) $77 - 65 - 6 = \boxed{}$

(c) $71 - 52 + 36 = \boxed{}$ (d) $75 + 24 - 23 = \boxed{}$

(e) $32 + 45 - 38 = \boxed{}$ (f) $81 - 45 + 36 = \boxed{}$

(g) $63 + 12 - 42 = \boxed{}$ (h) $53 - 37 + 15 = \boxed{}$

(i) $63 - 23 + 28 = \boxed{}$ (j) $26 + 49 - 12 = \boxed{}$

3 Fill in the boxes.

(a) $53 + \boxed{} = 70$ (b) $\boxed{} + 76 = 98$

(c) $\boxed{} - 72 = 7$ (d) $49 = 18 + \boxed{}$

(e) $5 = \boxed{} - 35$ (f) $\boxed{} + 18 = 18$

(g) $5 \times \boxed{} = 50$ (h) $8 \times 12 = \boxed{}$

(i) $32 \div \boxed{} = 8$ (j) $\boxed{} \div 2 = 7$

(k) $\boxed{} \times 10 = 4 \times 5$ (l) $36 \div 4 = \boxed{} \div 2$

4 Write the number sentence and answer the question.

(a)   There are 28 fewer ▲ than ⬤.

⬤⬤⬤ ... ⬤⬤   ▲ ▲ ... ▲

65 ⬤   How many ▲ are there?

Number sentence: _____

There are [   ] ▲.

(b) 16 pears were eaten.   How many were left?

There were 30 pears altogether.

Number sentence: _____

There were [   ] pears left.

(c)   24 sweets were eaten.   38 sweets were left.

How many sweets were there at first?

Number sentence: _____

There were [   ] sweets at first.

(d)   ?

Story books ⊢_____⌐_____⌐

Science books ⊢_____⌐   16

45

Number sentence: _____

There are [   ] story books.

(e)

Roses 37 bunches

Lilies

How many more?

72 bunches

Number sentence: _____

There are ▢ more bunches of lilies than roses.

(f) A bus can seat 40 people. A car can seat 4 people.
How many times as many people can a bus seat as a car?

Number sentence: _____

A bus can seat ▢ times as many people as a car.

(g) A 72-metre-long rope was cut into 9 equal pieces. How long is
each piece?

Number sentence: _____

Each piece is _____ long.

5 Write a suitable number in each box.

(a)
```
    3 ▢
  + ▢ 6
  ─────
    7 7
  ─────
```

(b)
```
    8 ▢
  - ▢ 5
  ─────
    4 0
  ─────
```

(c)
```
    ▢ 4
  + 3 ▢
  ─────
    6 3
  ─────
```

(d)
```
    ▢ 6
  - 1 ▢
  ─────
    4 9
  ─────
```

Chapter 2 Multiplication and division (II)

2.1 Multiplying and dividing by 7

 Learning objective Multiply and divide by 7

 Basic questions

1 Write the answers in words.

Two times seven is _____.

Four times seven is _____.

Seven times eight is _____.

One times seven is _____.

Seven times _____ is forty-nine.

_____ times seven is twenty-one.

_____ times _____ is sixty-three.

Ten times _____ is seventy.

Eleven times seven is _____.

_____ times seven is eighty-four.

2 Draw lines to help each cat catch the right fish.

Three times seven is twenty-one.

Seven times eight is fifty-six.

Six times seven is forty-two.

Five times seven is thirty-five.

$56 \div 7$

7×3

7×8

$35 \div 5$

5×7

$21 \div 7$

7×6

$42 \div 6$

3 Calculate.

(a) $1 \times 7 = \boxed{}$

(b) $7 \times 4 = \boxed{}$

(c) $35 \div 7 = \boxed{}$

(d) $63 \div 9 = \boxed{}$

(e) $0 \div 7 = \boxed{}$

(f) $9 \times 7 = \boxed{}$

(g) $14 \div 7 = \boxed{}$

(h) $56 \div 7 = \boxed{}$

(i) $7 \times 11 = \boxed{}$

(j) $12 \times 7 = \boxed{}$

(k) $84 \div 7 = \boxed{}$

(l) $70 \div 7 = \boxed{}$

(m) $49 = \boxed{} \times 7$

(n) $3 = \boxed{} \div 7$

(o) $77 = \boxed{} \times 7$

(p) $10 = \boxed{} \div 7$

4 Application problems.

(a) 14 sweets were shared equally by 7 children. How many sweets did each child get?

Answer: _____

(b) In one week, Zoe used 3 pieces of paper every day. How many pieces of paper did she use in the whole week?

Answer: _____

(c) In Maths lessons for Class 3A, the teacher divides the pupils into 7 groups. There are 5 pupils in each group. How many pupils are there altogether?

Answer: _____

(d) Class 3B has 6 more pupils than Class 3A. How many pupils are there in Class 3B?

Answer: _____

Challenge and extension question

5 Use the numbers below to make number sentences. Who can make the most number sentences? Operations can be used on both sides of the equation.

14, 42, 6, 7, 2, 4, 28, 35, 5

2.2 Multiplying and dividing by 3

 Learning objective Multiply and divide by 3

 Basic questions

1 Complete the multiplication facts. Then write multiplication and division sentences.

(a) Three times five is _____.

_____   _____

_____   _____

(b) _____ times _____ is thirty-three.

_____   _____

_____   _____

(c) Three times _____ is eighteen.

_____   _____

_____   _____

2 Calculate.

(a) $6 \times 3 = \boxed{}$ (b) $3 \times 8 = \boxed{}$ (c) $3 \times 9 = \boxed{}$

(d) $11 \times 3 = \boxed{}$ (e) $7 \times 3 = \boxed{}$ (f) $3 \times 12 = \boxed{}$

(g) $30 \div 3 = \boxed{}$ (h) $5 \times 3 = \boxed{}$ (i) $36 \div 3 = \boxed{}$

(j) $2 \times 9 = \boxed{} \times 3$ (k) $27 \div 9 = 3 \times \boxed{}$ (l) $4 \times \boxed{} = 2 \times 6$

(m) $3 \times \boxed{} = 12$ (n) $\boxed{} \times 3 = 12 + 3$ (o) $\boxed{} + 3 = 12 \times 3$

3 Complete the table.

Dividend	3		18	27		30	0
Divisor		7	6		3		3
Quotient	1	3		9	12	3	

4 Write a number sentence for each question.

(a) What is the product of 2 threes?

Number sentence: _____

(b) How many threes should be subtracted from 15 so the result is 0?

Number sentence: _____

(c) What is 7 times 3?

Number sentence: _____

5 Application problems.

(a) Joe and his parents visit a museum. The admission ticket is £8 per person. How much do they have to pay?

Answer: _____

Joe pays with a £50 note. How much change should he get?

Answer: _____

(b) 6 ducks are on a river. There are 4 times as many ducks on the bank as on the river. How many ducks are on the bank?

Answer: _____

How many ducks are there altogether?

Answer: _____

Challenge and extension questions

6 A bag of sweets can be divided equally into 3 groups. It can also be divided equally into 6 groups. What could be the smallest number of sweets in the bag?

Answer: _____

7 A monkey picked 24 peaches. She gave all the peaches to her two baby monkeys. The older baby monkey got 2 times as many peaches as the younger one.

(a) How many peaches did the younger baby monkey get?

Answer: _____

(b) How many peaches did the older monkey get?

Answer: _____

2.3 Multiplying and dividing by 6

Learning objective Multiply and divide by 6

Basic questions

1 Write two multiplication sentences in numbers and the multiplication fact in words for each picture.

(a)

Multiplication sentences in numbers:

_____

_____

Multiplication fact in words:

_____

(b)

Multiplication sentences in numbers:

_____

_____

Multiplication fact in words:

_____

(c)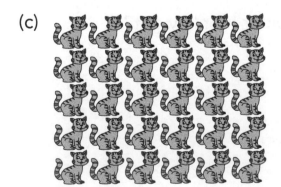

Multiplication sentences in numbers:

_____

_____

Multiplication fact in words:

_____

(d)

Multiplication sentences in numbers:

_____

_____

Multiplication fact in words:

_____

2 Write >, < or = in each ◯.

(a) 3×6 ◯ 9×2

(b) 6×12 ◯ 70

(c) 6×9 ◯ $5 \times 9 + 9$

(d) $60 \div 10$ ◯ $66 \div 6$

3 Fill in the boxes.

(a) $12 = \boxed{} \times \boxed{} = \boxed{} \times \boxed{} = \boxed{} \times \boxed{}$

(b) $36 = \boxed{} \times \boxed{} = \boxed{} \times \boxed{} = \boxed{} \times \boxed{}$

(c) $18 = \boxed{} \times \boxed{} = \boxed{} \times \boxed{} = \boxed{} \times \boxed{}$

4 Calculate.

(a) $6 \times 5 = \boxed{}$

(b) $7 \times 6 = \boxed{}$

(c) $9 \times 6 = \boxed{}$

(d) $24 \div 6 = \boxed{}$

(e) $3 \times 7 = \boxed{}$

(f) $8 \times 9 = \boxed{}$

(g) $0 \times 6 = \boxed{}$

(h) $36 \div 4 = \boxed{}$

(i) $3 \times 11 = \boxed{}$

(j) $6 \times 11 = \boxed{}$

(k) $12 \times 6 = \boxed{}$

(l) $66 \div 11 = \boxed{}$

(m) $6 \times 10 = \boxed{}$

(n) $72 \div 6 = \boxed{}$

(o) $6 = \boxed{} \div 6$

(p) $11 = \boxed{} \div 6$

5 Read the information and answer the questions below.

I worked 5 days this week for 8 hours each day.

Pete

I worked 6 days this week for 8 hours each day.

Jo

(a) How many hours did Jo work this week?

Answer: _____

(b) How many fewer hours did Pete work this week than Jo?

Answer: _____

 Challenge and extension questions

6 Given ● + ● + ● + ● = ▲ + ■ and ■ = ▲ + ▲,

if ● = 6, then ▲ = ☐ .

7 Simon starts reading his book at page 1. He reads 6 pages each day.

He reads the book for 4 days.

On the fifth day, Simon starts reading from page ☐ .

2.4 Multiplying and dividing by 9

Learning objective Multiply and divide by 9

Basic questions

1 Recall the multiplication facts, and write the answers in words and then in numbers.

(a) One times nine is _____.

$1 \times 9 =$ _____ $9 \times 1 =$ _____

(b) Three times nine is _____.

_____   _____

(c) Four times nine is _____.

_____   _____

(d) Eleven times nine is _____.

_____   _____

(e) _____ is fifty-four.

_____   _____

(f) _____ is eighteen.

_____   _____

2 What is the greatest number you can write in each box?

(a) $2 \times \boxed{} < 19$ (b) $10 \times \boxed{} < 25$ (c) $\boxed{} \times 9 < 23$

(d) $5 \times \boxed{} < 24$ (e) $\boxed{} \times 5 < 54$ (f) $\boxed{} \times 6 < 36$

(g) $8 \times \boxed{} < 20$ (h) $\boxed{} \times 9 < 55$ (i) $\boxed{} \times 9 < 100$

3 Choose three numbers from each group to write two multiplication sentences and two division sentences.

3, 9, 21, 7	6, 45, 5, 9	90, 9, 20, 10
_____	_____	_____
_____	_____	_____
_____	_____	_____
_____	_____	_____

4 Application problems.

(a) A rabbit collected 54 carrots and gave all of them equally to 6 baby rabbits. How many carrots did each baby rabbit get?

Answer: _____

(b) Suraj, Max and Erin went to Anya's home to celebrate her birthday. Anya's family prepared food including chocolate, cake and fruit.

The four children shared 36 chocolate bars equally. How many bars did each child get?

Answer: _____

(c) There were 9 bananas in a bunch. There were 3 bunches. How many bananas were there altogether?

Answer: _____

(d) There were 9 pieces of cake in each box. How many boxes were there for 18 pieces of cake?

Answer: _____

Challenge and extension questions

5 Complete the number patterns.

(a)

| 9 | 5 | 18 | 10 | 27 | 15 | ☐ | ☐ |

(b)

| 1 | 3 | 7 | 15 | 31 | ☐ | ☐ |

6 (a) It took Molly 18 seconds to run from the first floor to the third floor. How many seconds did it take Molly to run from one floor to the next floor on average?

Answer: _____

(b) How many seconds would it take Molly to run from the first floor to the eighth floor if she could keep up the same speed?

Answer: _____

2.5 Relationships between multiplications of 3, 6 and 9

Learning objective Use the relationship between the 3, 6 and 9 times tables

Basic questions

1 Group the objects and write the multiplication sentences for each picture.

(a)

3 in a group

☐ × ☐ = ☐

☐ × ☐ = ☐

(b)

6 in a group

☐ × ☐ = ☐

☐ × ☐ = ☐

(c)

9 in a group

☐ × ☐ = ☐

☐ × ☐ = ☐

2 Find the missing numbers.

(a) $18 = 3 \times \boxed{} = 6 \times \boxed{} = 9 \times \boxed{}$

(b) $36 = \boxed{} \times 4 = \boxed{} \times 6 = 3 \times \boxed{}$

(c) $54 = \boxed{} \times 9 = \boxed{} \times 6 = \boxed{} \times 3$

3 Write a number sentence for each question.

(a) What is the sum of adding 7 sixes together?

Number sentence: _____

(b) How many fours are there in 36?

Number sentence: _____

(c) How many threes should be subtracted from 24 so the result is 0? How many sixes?

Number sentence: _____

Number sentence: _____

4 Application problems.

(a) There are 9 roses. There are twice as many tulips as roses. How many tulips are there?

Answer: _____

(b) If a bouquet of flowers is made up of 9 flowers, how many bouquets can be made up with these tulips and roses?

Answer: _____

(c) There are 6 bunches of balloons and there are 6 balloons in each bunch. The balloons are to be shared equally among 3 children. How many balloons will each child get?

Answer: _____

Challenge and extension questions

5 There is a bag of marbles. They can be counted in both threes and sixes without any left over. What could be the smallest number of marbles in the bag?

Answer: _____

6 A goat is 3 times as heavy as a cat. The cat is 6 times as heavy as a squirrel. The squirrel is twice as heavy as a chick. How many times as heavy is the goat as the chick?

Answer: _____

2.6 Multiplication grid

Learning objective Explore numbers in a multiplication grid

Basic questions

1 Complete the multiplication grid.

									$10 \times 10 = 100$
								$9 \times 9 = 81$	
						$7 \times 7 = 49$			
				$5 \times 8 = 40$					
			$4 \times 5 = 20$						
		$3 \times 4 = 12$							
	$2 \times 2 = 4$								
$1 \times 1 = 1$									

2 Complete the multiplication facts. Then write multiplication and division sentences.

(a) Four times seven is _____.

_____   _____

_____   _____

(b) Five times eight is _____.

_____   _____

_____   _____

(c) Six times eleven is _____.

_____   _____

_____   _____

(d) Three times _____ is twelve.

_____   _____

_____   _____

(e) Four times _____ is twenty-four.

_____   _____

_____   _____

(f) _____ times nine is thirty-six.

_____   _____

_____   _____

3 Look at the pictures and write the number sentences.

(a)

Multiplication fact: _____

If we divide ☐ into ☐ equal groups, each group has ☐.

☐ ÷ ☐ = ☐

(b)

Multiplication fact: _____

There are ☐ groups of ☐ in ☐.

☐ ÷ ☐ = ☐

(c)

Multiplication fact: _____

Meaning: ☐ is ☐ times ☐.

☐ ÷ ☐ = ☐

(d) The doll costs ☐ times as much as the bear

Multiplication fact: _____

Meaning: ☐ times ☐ is ☐.

☐ ÷ ☐ = ☐

£9 £9 ?

Challenge and extension question

4 Write a suitable operation symbol in each ◯ so that the equation is correct.

(a) 2 ◯ 2 ◯ 2 ◯ 2 = 0   (b) 2 ◯ 2 ◯ 2 ◯ 2 = 1

(c) 2 ◯ 2 ◯ 2 ◯ 2 = 2   (d) 2 ◯ 2 ◯ 2 ◯ 2 = 3

2.7 Posing multiplication and division questions (1)

Learning objective Write multiplication and division number sentences

Basic questions

1 Look at each picture. Pose a question and then write a number sentence. The first one has been done for you.

(a)

How many sweets are there in total?

Number sentence:

  $5 \times 4 = 20$ (sweets)

(b)

Question: _____

_____

Number sentence: _____

_____

(c)

Question: _____

_____

Number sentence: _____

_____

(d)

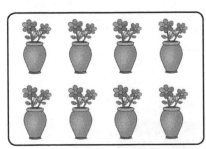

Question: _____

_____

Number sentence: _____

_____

2 Fill in the spaces by choosing the correct phrase or question below. Then write a number sentence and find the answer.

A school held a sports day. Year 3 pupils took part in various activities.

A 25 pupils took part in the rope skipping

B 16 pupils took part in a relay race

C How many pupils played table tennis?

D 42 pupils played football

 **E** How many times as many pupils played football as table tennis?

(a) _____ and there were 4 running tracks.

How many pupils ran on each running track?

Number sentence: _____

(b) _____ and each class sent 5 pupils to take part in the rope skipping. How many classes were there in Year 3?

Number sentence: _____

(c) _____, which was 7 times as many as those

playing table tennis. _____

Number sentence: _____

Challenge and extension questions

3 (a) In a long distance race, Alvin was 60 metres ahead of Maya. Suraj was 100 metres behind Holly. Maya was 20 metres behind Suraj. Who was the first runner in the race?

Answer: _____

(b) How many metres is the difference between the first runner and the last runner?

Answer: _____

4 Orla's mum bought 16 apples. She gave half of the apples to Grandma. Then she gave half of the rest to Orla. How many apples did Mum still have?

Answer: _____

2.8 Posing multiplication and division questions (2)

Learning objective Write multiplication and division number sentences

Basic questions

1 Read the text below. Some facts can be linked to write questions that can be solved using multiplication and division. The first one has been done for you.

- These are the flowers in Avni's garden.
- There are 10 pots of roses and each pot contains 4 roses.
- There are 36 lilies.
- There are 9 pots of tulips with each pot containing 2 tulips, and there are 72 orchids planted in 9 pots.
- There are also 6 empty pots.

(a) Linked fact(s): *There are 10 pots of roses and each pot contains 4 roses.*

Question: *How many roses are there in total?*

Number sentence: *10 × 4 = 40*

(b) Linked fact(s): _____

Question: _____

Number sentence: _____

(c) Linked fact(s): _____

Question: _____

Number sentence: _____

2 Application problems.

(a) Ethan has made 10 origami cranes. Ben, Sofia and Abena have each made as many origami cranes as Ethan. How many origami cranes have they made altogether?

Answer: _____

(b) 9 metres was cut from a ribbon. It is now 5 times as long as the piece that was cut off. How long is the remaining piece?

Answer: _____

How long was the original ribbon?

Answer: _____

(c) May and her parents visited the cinema over the weekend. The ticket price was £10 per person. How much did they pay?

Answer: _____

(d) There were 6 birds in each tree. How many birds were there in 5 trees?

Answer: _____

After 18 birds flew away, how many birds were left?

Answer: _____

Challenge and extension questions

3 It took Finn 60 seconds to walk from the first floor to the third floor. How long would it take him to walk from the first floor to the sixth floor at the same speed?

Answer: _____

4 Fold a 16-metre rope in half. Then fold the rope again so that you have four pieces. How long is each piece?

Answer: _____

2.9 Using multiplication and addition to express a number

 Learning objective Solve problems involving multiplication and addition

 Basic questions

1 Look at the pictures and fill in the answers.

(a)

There are [ ] 🥕 altogether.

Method I used: _____

Another method is: _____

(b)

There are [ ] 🥬 altogether.

Method I used: _____

Another method is: _____

2 What is the greatest number you can write in each box?

(a) $6 \times \boxed{} < 45$

(b) $7 \times \boxed{} < 40 - 5$

(c) $27 + 8 > 6 \times \boxed{}$

(d) $\boxed{} \times 9 < 32$

(e) $\boxed{} \times 5 < 4 \times 7$

(f) $8 \times \boxed{} < 20 + 27$

(g) $22 = \boxed{} \times 5 + \boxed{}$

(h) $63 = \boxed{} \times 8 + \boxed{}$

(i) $53 = \boxed{} \times 5 + \boxed{}$

(j) $21 = \boxed{} \times 4 + \boxed{}$

(k) $69 = \boxed{} \times 7 + \boxed{}$

(l) $25 = \boxed{} \times 6 + \boxed{}$

(m) $59 = \boxed{} \times 9 + \boxed{}$

(n) $67 = \boxed{} \times 7 + \boxed{}$

(o) $47 = \boxed{} \times 4 + \boxed{}$

(p) $79 = \boxed{} \times 8 + \boxed{}$

3 Fill in the boxes.

(a) $19 = 2 \times \boxed{} + \boxed{}$

(b) $32 = 3 \times \boxed{} + \boxed{}$

(c) $19 = 3 \times \boxed{} + \boxed{}$

(d) $32 = 4 \times \boxed{} + \boxed{}$

(e) $19 = 4 \times \boxed{} + \boxed{}$

(f) $32 = 5 \times \boxed{} + \boxed{}$

(g) $19 = 5 \times \boxed{} + \boxed{}$

(h) $32 = 6 \times \boxed{} + \boxed{}$

(i) $19 = 6 \times \boxed{} + \boxed{}$

(j) $32 = 7 \times \boxed{} + \boxed{}$

(k) $19 = 7 \times \boxed{} + \boxed{}$

(l) $32 = 8 \times \boxed{} + \boxed{}$

(m) $19 = 8 \times \boxed{} + \boxed{}$

(n) $32 = 9 \times \boxed{} + \boxed{}$

(o) $19 = 9 \times \boxed{} + \boxed{}$

(p) $32 = 10 \times \boxed{} + \boxed{}$

(q) $19 = 10 \times \boxed{} + \boxed{}$

(r) $32 = 11 \times \boxed{} + \boxed{}$

(s) $19 = 11 \times \boxed{} + \boxed{}$

(t) $32 = 12 \times \boxed{} + \boxed{}$

4 Use the numbers below to write number sentences with both multiplication and addition.

15 26 6 2 4 3 7 14

Here is an example: 6 × 4 + 2 = 26

_____   _____

_____   _____

_____   _____

 Challenge and extension question

5 A box of biscuits contains fewer than 40 biscuits. It is exactly enough to share equally with 6 children. If it is shared equally with 7 children, there is 1 biscuit left over.

There are ⬚ biscuits in the box.

2.10 Division with a remainder

Learning objective Understand remainders in division as leftovers

Basic questions

1 Group the objects equally and write number sentences.
There are 22 pencils in total.

(a)

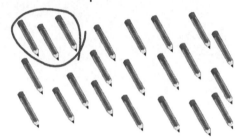

There are ☐ groups.

There is/are ☐ left over.

Number sentence:

_____

(b)

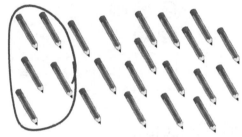

There are ☐ groups.

There is/are ☐ left over.

Number sentence:

_____

2 Put the flowers into the 4 vases equally.

Each vase has ☐ flowers. There are ☐ flowers left over.

Number sentence: _____

3 Complete the division sentences to match each representation.

(a)

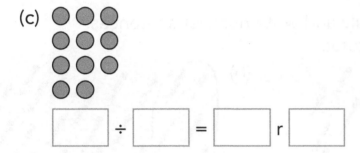

$9 \div 4 = \boxed{} \text{ r } \boxed{}$

(b)

$\boxed{} \div 5 = \boxed{} \text{ r } \boxed{}$

(c)

$\boxed{} \div \boxed{} = \boxed{} \text{ r } \boxed{}$

4 Calculate.

(a) $17 \div 3 = \boxed{} \text{ r } \boxed{}$ (b) $23 \div 7 = \boxed{} \text{ r } \boxed{}$

(c) $38 \div 6 = \boxed{} \text{ r } \boxed{}$ (d) $31 \div 4 = \boxed{} \text{ r } \boxed{}$

(e) $64 \div 9 = \boxed{} \text{ r } \boxed{}$ (f) $47 \div 8 = \boxed{} \text{ r } \boxed{}$

(g) $41 \div 7 = \boxed{} \text{ r } \boxed{}$ (h) $55 \div 8 = \boxed{} \text{ r } \boxed{}$

(i) $92 \div 10 = \boxed{} \text{ r } \boxed{}$ (j) $58 \div 9 = \boxed{} \text{ r } \boxed{}$

5 Application problems.

(a) Mr Bruce shared a box of water equally between Kit, Marlon, Ava and Amina. There were 9 bottles of water in the box. How many bottles did each child get?

Answer: _____

How many were left over?

Answer: _____

(b) A monkey gave her 7 baby monkeys some peaches. Each baby monkey got 6 peaches. There were 5 peaches left over. How many peaches were there altogether?

Answer: _____

(c) Mahmud has 50 | to make ⬡ without sharing edges. How many ⬡ can he make? How many | will be left over?

Answer: _____

Challenge and extension question

6 All the pupils in Class 3G are grouped to take part in a school activity. If they are grouped in sevens, there is one pupil left over. If they are grouped in sixes, there is also one pupil left over. Given that there are fewer than 50 pupils in the class, the number of pupils in the class is ☐ .

2.11 Calculation of division with a remainder (1)

Learning objective Solve division problems with remainders

Basic questions

1 Each box contains four pieces of .

(a) There are ☐ pieces of ◠ altogether.

(b) They are shared by ☐ children equally and each child gets ☐ pieces.

(c) There are ☐ pieces left over.

(d) Number sentence: _____

2 There are 45 🍎.

(a) 6 🍎 are put onto each plate. It is enough to fill ☐ plates.

(b) There are ☐ 🍎 left over.

(c) Number sentence: _____

3 What is the greatest number you can write in each circle?

(a) $7 \times \bigcirc < 41$

(b) $6 \times \bigcirc < 35$

(c) $3 \times \bigcirc < 26$

(d) $9 \times \bigcirc < 38$

(e) $\bigcirc \times 5 < 43$

(f) $\bigcirc \times 2 < 20$

(g) $\bigcirc \times 7 < 32$

(h) $4 \times \bigcirc < 43$

(i) $\bigcirc \times 8 < 65$

4 Are these calculations correct? Put a tick (✓) for yes and a cross (✗) for no in each box.

(a) $46 \div 8 = 5 \text{ r } 6$ ☐

(b) $36 \div 6 = 5 \text{ r } 6$ ☐

(c) $55 \div 7 = 7 \text{ r } 6$ ☐

(d) $44 \div 5 = 8 \text{ r } 4$ ☐

(e) $80 \div 12 = 7 \text{ r } 4$ ☐

(f) $99 \div 10 = 9 \text{ r } 9$ ☐

5 Write a number sentence for each question.

(a) There are 31 days in May and 7 days in a week. How many weeks plus how many days are there in May?

Number sentence: _____

(b) If each child gets 6 sweets, then how many children can share 50 sweets? How many sweets are left over?

Number sentence: _____

(c) A tailor makes coats with 5 buttons. How many coats will 49 buttons go on? How many buttons will be left over?

Number sentence: _____

Challenge and extension question

6 A class of 26 pupils plans to go boating.

Small boat – 4 people

Big boat – 6 people

(a) If all the pupils take small boats, how many small boats will they need?

Answer:

(b) If all the pupils take big boats, how many big boats will they need?

Answer:

(c) Can you offer some other suggestions for renting the boats?

Answer: _____

_____

2.12 Calculation of division with a remainder (2)

Learning objective Solve division problems with remainders

Basic questions

1 Calculate mentally.

(a) $37 \div 9 =$ ☐   (b) $47 \div 5 =$ ☐   (c) $43 \div 8 =$ ☐

(d) $19 \div 7 =$ ☐   (e) $69 \div 8 =$ ☐   (f) $62 \div 7 =$ ☐

(g) $26 \div 3 =$ ☐   (h) $27 \div 6 =$ ☐   (i) $9 \div 2 =$ ☐

(j) $14 \div 4 =$ ☐   (k) $95 \div 10 =$ ☐   (l) $63 \div 6 =$ ☐

2 Fill in the boxes.

(a) ☐ $\div 6 = 4$ r 4 (b) ☐ $\div 7 = 6$ r 2 (c) ☐ $\div 5 = 9$ r 1

(d) ☐ $\div 9 = 3$ r 8 (e) ☐ $\div 3 = 7$ r 2 (f) ☐ $\div 7 = 7$ r 4

3 Find the answers.

(a) Maria's plates are big enough to fit 6 apples.

15 apples can be put on ☐ plates with ☐ apples left over.

Number sentence: ☐ $\div$ ☐ $=$ ☐ r ☐

(b) Some sweets were shared equally between 8 children. After each child got 6 sweets, there were 4 sweets left over.

How many sweets were there altogether?

Number sentence: ☐ × ☐ + ☐ = ☐

(c) 24 bananas were given to 5 monkeys equally, and each monkey got ☐ bananas, with ☐ bananas left over.

Number sentence: ☐ ÷ ☐ = ☐ r ☐

(d) In ☐ ÷ 6 = 7 r ☐ , the remainder could be ☐ .
The greatest possible remainder is ☐ .
When the remainder is the greatest, the dividend is ☐ .

(e) From the number sentence 5 × 7 + 5 = 40, we can write another number sentence:

40 ÷ ☐ = ☐ r ☐ .

4 Use the number sentences of multiplication and addition to write number sentences of division with remainders.

(a) 4 × 3 + 1 = 13

13 ÷ ☐ = 4 r ☐

13 ÷ ☐ = 3 r ☐

(b) 4 × 6 + 2 = 26

☐ ÷ ☐ = ☐ r ☐

☐ ÷ ☐ = ☐ r ☐

5 Application problems.

(a) In a flower shop, a bouquet should have 6 flowers. How many bouquets can be made with 25 flowers? How many flowers will be left over?

Answer: _____

(b) 58 oranges are to be shared among 7 people equally. How many oranges will each person get? How many oranges will be left over?

Answer: _____

If each person should get 9 oranges, how many more oranges are needed?

Answer: _____

Challenge and extension question

6 In the list of numbers 1, 3, 5, 1, 3, 5, 1, 3, 5, …,

the 26th number is ⬚.

The sum of these 26 numbers is ⬚.

2.13 Calculation of division with a remainder (3)

 Learning objective Solve division problems with remainders

 Basic questions

1 Calculate mentally.

(a) $5 \times 4 = \boxed{}$

(b) $48 \div 6 = \boxed{}$

(c) $28 \div 4 = \boxed{}$

(d) $6 \times 9 = \boxed{}$

(e) $16 \div 4 = \boxed{}$

(f) $58 \div 8 = \boxed{}$

(g) $4 \times 8 = \boxed{}$

(h) $56 - 8 = \boxed{}$

(i) $17 \div \boxed{} = 8 \text{ r } 1$

(j) $\boxed{} \div 9 = 4 \text{ r } 2$

(k) $\boxed{} \div 6 = 8 \text{ r } 2$

(l) $\boxed{} \div 8 = 9 \text{ r } 7$

(m) $\boxed{} \div 7 = 4 \text{ r } 5$

(n) $\boxed{} \div 10 = 7 \text{ r } 1$

2 Find the remainder from each set of numbers and then work out the dividend.

(a) ⑤⑥⑦⑧

$\boxed{} \div 6 = 6 \text{ r } \boxed{}$

(b) ④⑤⑥⑦

$\boxed{} \div 5 = 3 \text{ r } \boxed{}$

(c) ③④⑤⑥

$\boxed{} \div 4 = 7 \text{ r } \boxed{}$

(d) ①②③④

$\boxed{} \div 2 = 10 \text{ r } \boxed{}$

3 Complete each calculation and find the answers.

(a) $\bigcirc \div 4 = 4$ r $\boxed{}$

The greatest possible number in the $\bigcirc$ is $\boxed{}$.

The smallest possible number is $\boxed{}$.

(b) $\bigcirc \div 9 = 3$ r $\boxed{}$

The greatest possible number in the $\bigcirc$ is $\boxed{}$.

The smallest possible number is $\boxed{}$.

4 Application problems.

(a) Jo's family has raised 3 ducks and 4 times as many chicks as ducks. How many chicks has the family raised?

Answer: _____

(b) The family has also raised 23 rabbits. A hutch can house 4 rabbits. How many hutches does the family need to house all the rabbits?

Answer: _____

(c) Jo has £50 to buy some hamsters for her family. Each hamster costs £9. How many hamsters can she buy?

Answer: _____

(d) The family has 5 black goldfish and 13 red goldfish.

The number of red goldfish is $\boxed{}$ more than $\boxed{}$ times the number of black goldfish.

Challenge and extension questions

5 Fill in the boxes.

(a) 35 ÷ ☐ = ☐ r 5   (b) 57 ÷ ☐ = ☐ r 3

(c) 35 ÷ ☐ = ☐ r 3   (d) 57 ÷ ☐ = ☐ r 1

6 Six children are sharing some stickers. If 5 more stickers are added, then each child can have 5 stickers.

There are ☐ stickers.

Chapter 2 test

1 Calculate mentally.

(a) $7 \times 7 =$ ☐

(b) $8 + 15 =$ ☐

(c) $6 \times 11 =$ ☐

(d) $36 \div 3 =$ ☐

(e) $35 + 5 =$ ☐

(f) $72 \div 9 =$ ☐

(g) $34 \div 8 =$ ☐

(h) $88 \div 8 =$ ☐

(i) $20 \div 5 =$ ☐

(j) $0 \div 10 =$ ☐

(k) $49 \div 7 =$ ☐

(l) $100 - 36 + 36 =$ ☐

(m) $27 =$ ☐ $- 50$

(n) $19 =$ ☐ $\times 2 +$ ☐

(o) $16 =$ ☐ $\times 6 +$ ☐

(p) $37 =$ ☐ $\times 8 +$ ☐

2 Use the column method to calculate.

(a) $33 + 38 =$ ☐

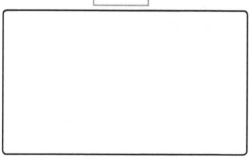

(b) $65 - 38 =$ ☐

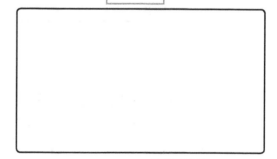

(c) $64 + 36 - 17 =$ ☐

(d) $85 - 67 + 39 =$ ☐

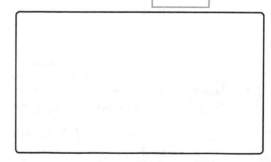

3 Write each number sentence.

(a) 45 is 29 more than what number?

Number sentence: _____

(b) What is the result of adding 12 twos together?

Number sentence: _____

(c) What is the result of multiplying 3 threes?

Number sentence: _____

(d) In a division calculation, the quotient and the divisor are both 8 and the remainder is 3. What is the dividend?

Number sentence: _____

4 Application problems.

(a) Look at each picture, write a number sentence and calculate.

(i)

Number sentence:

_____

?

(ii)

Pencil ⊢———4———⊣

Pen ⊢————┼————┼————⊣

?

Number sentence:

_____

(iii)

19 baskets taken   23 baskets left

⊢————┼————⊣

How many baskets were there at first?

Number sentence:

_____

(b) (i) 8 cartons of milk will fill up a box. How many boxes will 50 cartons of milk fill?

Answer: _____

(ii) How many cartons will be left over?

Answer: _____

(c) Katy and her parents went on holiday for 3 weeks. For how many days were they on holiday?

Answer: _____

(d) Noah bought some flowers. He put 8 flowers in 1 vase. After he filled 4 vases, there were 4 flowers left over. How many flowers did Noah buy?

Answer: _____

(e) There were 72 cupcakes in a bakery. The baker put 9 cupcakes in 1 box. She packed 5 boxes. How many cupcakes did she put into boxes? How many cupcakes were still left unpacked?

Answer: _____

5 Understand concepts.

(a) Look at each picture and fill in the missing numbers.

(i)

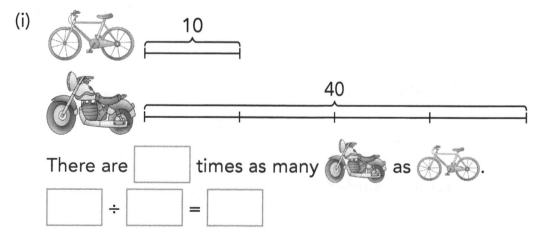

There are ☐ times as many 🏍 as 🚲.

☐ ÷ ☐ = ☐

(ii)

□ ◯ □ = □ groups

Meaning: □ contains □ groups of □.

□ ◯ □ = □ watermelons

Meaning: dividing □ into □ equal groups, each group has □.

(iii)

Each pair has 2 gloves. There are □ pairs.

How many gloves are there altogether?

□ ◯ □ = □

1 pair of costs £6.

Can you write a question based on the information?

Can you solve it?

Question: _____

_____

Number sentence: _____

(b) Fill in the boxes.

  (i) What is the greatest number you can write in each box?

    $4 \times \boxed{} < 21$ $\boxed{} \times 5 < 46$

    $\boxed{} \times 9 < 98$ $52 - \boxed{} > 40$

  (ii) Write the correct number in each box.

    $15 \times 2 = 5 \times \boxed{}$ $12 \times \boxed{} = \boxed{} \times 24$

    $2 + 2 + 2 + 4 = \boxed{} \times \boxed{}$

  (iii) Calculate smartly.

    $43 - 39 = \boxed{} - \boxed{} = \boxed{}$

  (iv) In a division calculation, the remainder should be

    _____ than the divisor.

    When a number is divided by 4, its quotient is 9 and the

    greatest possible remainder is $\boxed{}$.

  (v) When a number is divided by 9, the quotient is 4 and the
       remainder is 3.

    The number is $\boxed{}$.

  (vi) A pack contains fewer than 30 jellybeans. The jellybeans are
        exactly enough for 5 children to share equally. If they are
        shared equally by 6 children, there is 1 jellybean left over.

    There are $\boxed{}$ jellybeans in the pack.

(c) True or false? (Put a ✓ for true and a ✗ for false in each box.)

  (i) $33 \div 7 = 5 \text{ r } 2$ . . . . $\boxed{}$

  (ii) If $\triangle \div 6 = \bigcirc \text{ r } \blacksquare$, then the greatest
        possible number that $\blacksquare$ can be is 5 . . . . $\boxed{}$

  (iii) $12 \div 0 = 0$. . . . . $\boxed{}$

(d) Multiple choice questions. (For each question, choose the correct answer and write the letter in the box.)

(i) $13 \div 3 = 4$ r 1 is read as ☐.

A. 13 dividing by 3 equals 4.

B. 13 divided by 3 equals 4 with a remainder of 1.

C. 13 divided by 3 equals 4.

D. 13 dividing by 3 equals 4 with a remainder of 1.

(ii) The quotient is 10, the divisor is 5, and the dividend is 50. The number sentence is ☐.

A. $5 \times 10 = 50$	**B.** $10 \times 5 = 50$
C. $50 \div 5 = 10$	**D.** $50 \div 10 = 5$

(iii) The number sentence with the same answer as 8×2 is ☐.

A. $8 + 2$ **B.** 3×6 **C.** 2×9 **D.** 4×4

Chapter 3 Knowing numbers up to 1000

3.1 Knowing numbers up to 1000 (1)

 Learning objective Read, write and partition numbers up to 1000

 Basic questions

1 Calculate mentally.

(a) $7 \times 7 = \boxed{}$ (b) $80 \div 10 = \boxed{}$ (c) $45 - 18 = \boxed{}$

(d) $50 \div 9 = \boxed{}$ (e) $74 + 6 = \boxed{}$ (f) $5 \times 8 = \boxed{}$

(g) $44 + 13 = \boxed{}$ (h) $54 \div 5 = \boxed{}$ (i) $0 \div 3 = \boxed{}$

(j) $10 \times 10 = \boxed{}$ (k) $70 - 7 = \boxed{}$ (l) $63 \div 9 = \boxed{}$

2 Look at each diagram and write the number. The first one has been done for you.

(a) (i)

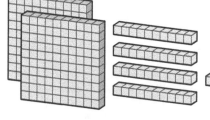

In numerals: 243

In words: two hundred and forty-three

(ii)

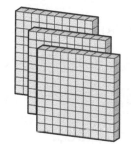

In numerals: _____

In words: _____

(iii)

In numerals: _____

In words: _____

(iv)

In numerals: _____

In words: _____

(v)

In numerals: _____

In words: _____

(b) (i)

Hundreds	Tens	Ones
6	0	5

In numerals: _____

In words: _____

(ii)

Hundreds	Tens	Ones
8	2	4

In numerals: _____

In words: _____

3 Complete the sentences.

(a) 856 is made up of ☐ hundreds, ☐ tens and ☐ ones.

(b) Counting from the right in a 4-digit number, the first place is

the _____ place, the tens place is the _____ place

and the thousands place is the _____ place.

(c) 707 is written in words as _____.

_____

7 in the ones place means ☐ ones.

7 in the hundreds place means ☐ hundreds.

The difference between them is ☐.

(d) 4 hundreds and 3 ones make ☐.

It is written in words as _____.

_____

4 Fill in the boxes.

(a) 462 = ☐ + ☐ + ☐

(b) 1050 = ☐ + ☐ + ☐ + ☐

(c) 788 = ☐ + ☐ + ☐

(d) 300 + 90 + 0 = ☐

(e) 800 + 8 = ☐

Challenge and extension questions

5 1 🦢 = 3 🐟   1 🐟 = 10 🥚   2 🦢 = ☐ 🥚

6 1 jug of water = 2 bottles of water

1 bottle of water = 4 cups of water

2 bottles of water = ☐ cups of water

1 jug of water = ☐ cups of water

3.2 Knowing numbers up to 1000 (2)

Learning objective Place value of numbers up to 1000

Basic questions

1 Calculate mentally.

(a) $4 \times 7 \times 2 =$

(b) $51 - 30 + 5 =$

(c) $45 + 91 + 29 =$

(d) $60 \div 6 + 29 =$

(e) $4 \times 3 - 0 =$

(f) $0 \times 4 + 62 =$

(g) $51 + 37 - 59 =$

(h) $1 \times 8 - 7 =$

(i) $3 \times 0 + 48 =$

(j) $16 + 75 - 41 =$

2 Look at the diagrams and then write the numbers in numerals.

(a)

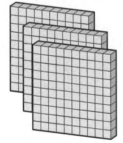

Written as:

(b)

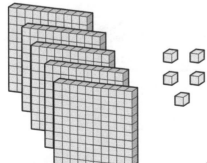

Written as:

3 Read the numbers and then write them in words or in numerals.

    (a) 635 in words: _____

    _____

    (b) 302 in words: _____

    _____

    (c) Nine hundred and thirty-six in numerals: ☐

    (d) 1000 in words: _____

    (e) Four hundred in numerals: ☐

4 Fill in the boxes.

    (a) Four hundred and eight is written in numerals as ☐.

       It is a ☐ -digit number.

       It consists of ☐ hundreds and ☐ ones.

    (b) The number consisting of 6 tens and 4 hundreds is ☐.

    (c) 10 one hundreds is ☐ . ☐ one hundreds is 500.

    (d) The numbers that come before and after 300 are ☐ and ☐ respectively.

    (e) There are ☐ tens in 470.

5 Draw diagrams to represent numbers and then write them in numerals. The first one has been done for you.

One hundred and eighty-seven

Draw:

Written as: 187

(a) Six hundred and six

Draw:

Written as:

(b) Two hundred and eighty

Draw:

Written as:

Challenge and extension question

6 The numbers 4, 0 and 2 can be used to make ☐ 3-digit numbers. Write these numbers and put them in order starting from the greatest, using > to link them.

Answer: _____

3.3 Number lines (to 1000) (1)

 Learning objective Compare and order numbers up to 1000

 Basic questions

1 Complete these questions about number lines.

(a) Mark these numbers on the number line.

A = 540   B = 780   C = 810   D = 600   E = 450   F = 370

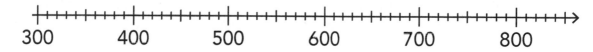

```
+++++++++++++++++++++++++++++++++++++++++++++++++>
300      400       500       600       700       800
```

(b) Find the number that each letter stands for.

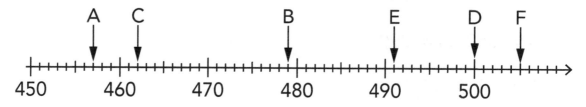

```
      A   C              B        E      D   F
      ↓   ↓              ↓        ↓      ↓   ↓
+++++++++++++++++++++++++++++++++++++++++++++++++>
450       460       470       480       490       500
```

A = ☐   B = ☐   C = ☐

D = ☐   E = ☐   F = ☐

2 Fill in the boxes with numbers based on the given information.

(a) Write the numbers that come before and after each number.

(i) ☐ , 278, ☐

(ii) ☐ , 999, ☐

(iii) ☐ , 406, ☐

(b) Write the tens numbers that come before and after each number.

(i) ⬚ , 390, ⬚

(ii) ⬚ , 455, ⬚

(iii) ⬚ , 789, ⬚

(c) Write the hundreds numbers that come before and after each number.

(i) ⬚ , 657, ⬚

(ii) ⬚ , 405, ⬚

(iii) ⬚ , 790, ⬚

3 Count and complete the number patterns.

(a) 567, 568, 569, ⬚ , ⬚

(b) 350, 370, 390, ⬚ , ⬚ , 450

(c) 743, 742, 741, ⬚ , ⬚

(d) 250, ⬚ , ⬚ , 550, 650, 750

4 Choose suitable numbers from the list below to answer these questions.

| 439 | 501 | 92 | 888 | 654 | 499 | 328 | 1000 |

(a) The numbers greater than 400 but less than 500 are:

_____

(b) The number that is 111 less than 999 is:

_____

(c) The numbers that come before and after 500 are:

_____

(d) Write the above numbers in order, starting from the greatest.

_____

5 Write the numbers.

(a) Write the numbers greater than 498 but less than 505.

_____

(b) Write the hundreds numbers less than 900.

_____

(c) Write all the 3-digit numbers that are less than 200 and have the same digit in the ones place and in the tens place.

_____

Challenge and extension questions

6 In some 3-digit numbers, the sum of the three digits is 15 and the digit in the hundreds place is twice the digit in the ones place.

These 3-digit numbers are: _____ .

7 When you write numbers from 200 to 300, you need to write the digits:

1 ☐ times, 2 ☐ times and 0 ☐ times.

3.4 Number lines (to 1000) (2)

Learning objective Compare and order numbers up to 1000

Basic questions

1 Complete these questions about number lines.

560   570   580   590   600   610

(a) Mark the following numbers on the number line.

     A = 587   B = 565   C = 599

     D = 571   E = 602   F = 618

(b) Fill in the boxes.

     (i) A + ☐ = 590   (ii) B + ☐ = 570   (iii) C − ☐ = 590

     (iv) D − ☐ = 570   (v) E + ☐ = 610   (vi) F − ☐ = 610

(c) Fill in the boxes.

     (i) A + ☐ = 600   (ii) B − ☐ = 500   (iii) C + ☐ = 600

     (iv) D − ☐ = 500   (v) E + ☐ = 700   (vi) F − ☐ = 600

(d) Put the six numbers A, B, C, D, E and F in order, starting with the smallest.

     ☐ < ☐ < ☐ < ☐ < ☐ < ☐

2 Count and complete the number patterns.

(a) ☐ , ☐ 290, 285, ☐ , ☐

(b) 486, 488, ☐ , ☐

(c) 123, 223, 323, ☐ , ☐ , ☐

3 Write the numbers in order.

(a) Start with the greatest: 175   715   517   157   751   117

_____

(b) Start with the smallest: 869   886   689   668   969   898

_____

(c) From the greatest to the smallest: all the 3-digit numbers with 7 in the ones place and 6 in the tens place.

_____

4 Write >, < or = in each ◯.

(a) 428 ◯ 482   (b) 789 ◯ 787   (c) 543 ◯ 453

(d) 603 ◯ 630   (e) 1000 ◯ 999   (f) 135 ◯ 125

(g) 155 ◯ 205   (h) 299 ◯ 301

(i) 438 ◯ 400 + 30 + 8

5 What is the greatest number you can write in each box?

(a) ☐ 65 < 655   (b) 7 ☐ 8 > 778   (c) 453 < 4 ☐ 3

(d) 321 > ☐ 21   (e) 642 > ☐ 43   (f) 795 < 79 ☐

Challenge and extension questions

6 Fill in the boxes with the same number so the subtraction is correct.

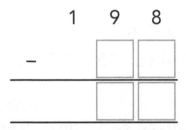

7 Look at this calculation.

$70 \div \boxed{} = \bigcirc \text{ r } 6$

Write a suitable 1-digit number in each of the $\boxed{}$ and $\bigcirc$ so that the number sentence is correct.

Write out the number sentence(s).

_____

_____

3.5 Fun with the place value chart (1)

 Learning objective Place value of numbers up to 1000

 Basic questions

1 Look at the diagrams and write the numbers.

(a)

Hundreds	Tens	Ones
●●●● ●●●	●●●	●●●●

In numerals: _____

In words: _____

_____

(b)

Hundreds	Tens	Ones
●●●●	●●●	

In numerals: _____

In words: _____

_____

(c)

Hundreds	Tens	Ones
●●●		●

In numerals: _____

In words: _____

_____

(d)

Hundreds	Tens	Ones
●●●●		

In numerals: _____

In words: _____

_____

2 Draw dots in the place value chart to represent each number given.

(a) 507

Hundreds	Tens	Ones

(b) 800

Hundreds	Tens	Ones

3 Meera has drawn the dots representing 264 in the place value chart below. Joe adds one more dot. What could the new number be?

Hundreds	Tens	Ones
● ●	●●● ●●●	●●●●

Hundreds	Tens	Ones

In numerals: _____

In words: _____

_____

Hundreds	Tens	Ones

In numerals: _____

In words: _____

_____

Hundreds	Tens	Ones

In numerals: _____

In words: _____

_____

 ## Challenge and extension questions

4 Hai folded a piece of paper in half three times. He then drew a flower in the centre and cut it out.

Hai cut out ☐ flowers in the paper.

5 After drinking half a cup of milk, Asha filled up the cup with water. She then drank half of the contents of the cup again. After that, she filled up the cup with water a second time. She then drank the whole of the contents of the cup.

In total, Asha drank ☐ cup(s) of milk and ☐ cup(s) of water.

3.6 Fun with the place value chart (2)

 Learning objective Place value of numbers up to 1000

 Basic questions

1 Draw three dots in each place value chart to represent six different 3-digit numbers.

(a)

Hundreds	Tens	Ones

In numerals: _____

In words: _____

_____

(b)

Hundreds	Tens	Ones

In numerals: _____

In words: _____

_____

(c)

Hundreds	Tens	Ones

In numerals: _____

In words: _____

_____

(d)

Hundreds	Tens	Ones

In numerals: _____

In words: _____

_____

(e)

Hundreds	Tens	Ones

In numerals: _____

In words: _____

_____

(f)

Hundreds	Tens	Ones

In numerals: _____

In words: _____

_____

2 Draw dots in the first place value chart to represent 153.

Then, in each place value chart below, move one dot into another column and write the number.

Hundreds	Tens	Ones

(a)

Hundreds	Tens	Ones

In numerals: _____

In words: _____

_____

(b)

Hundreds	Tens	Ones

In numerals: _____

In words: _____

_____

(c)

Hundreds	Tens	Ones

In numerals: _____

In words: _____

_____

(d)

Hundreds	Tens	Ones

In numerals: _____

In words: _____

_____

(e)

Hundreds	Tens	Ones

In numerals: _____

In words: _____

_____

(f)

Hundreds	Tens	Ones

In numerals: _____

In words: _____

_____

3 Write in the correct numbers.

(a) 643 = ☐ + ☐ + ☐

(b) 300 + 30 + 8 = ☐

(c) 302 = ☐ + ☐ + ☐

(d) 900 + 0 + 9 = ☐

Challenge and extension questions

4 A group of children form a line for a running race. Lily is in the twelfth place counting from the start. Ellis is in the twelfth place counting from the end. Lily is exactly in front of Ellis.

There are ☐ children in total.

5 A book has 40 pages. If a bookmark is put in every three pages starting from the first page, it will have ☐ bookmarks in total.

Chapter 3 test

1 Calculate mentally.

(a) $7 \times 5 =$ ☐

(b) $1000 - 960 =$ ☐

(c) $3 \times 9 =$ ☐

(d) $72 \div 8 =$ ☐

(e) $45 \div 9 =$ ☐

(f) $4 \times 8 =$ ☐

(g) $640 + 40 =$ ☐

(h) $43 - 12 =$ ☐

(i) $860 - 50 =$ ☐

(j) $55 + 24 =$ ☐

(k) $83 - 8 =$ ☐

(l) $370 - 300 =$ ☐

2 Write the numbers in words and do a drawing to represent them.

(a) 165 In words: _____

Drawing:

(b) 608 In words: _____

Drawing:

3 Look at the diagrams and write the numbers.

(a)

Thousands	Hundreds	Tens	Ones
●			

In numerals: _____

In words: _____

(b)

In numerals: _____

In words: _____

4 Write a number sentence for each question.

(a) How much more is the sum of 8 twos than 10?

Number sentence: _____

(b) Add 27 to the difference between 31 and 15. How much is it?

Number sentence: _____

(c) Both addends are 7. What is the sum?

Number sentence: _____

5 Fill in the boxes.

(a) The even numbers that come before and after 700 are [ ]
and [ ] .

(b) Complete these questions about number lines.

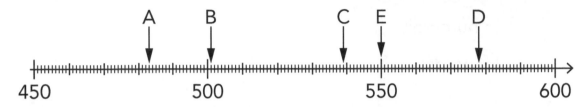

(i) Mark 456, 580, 489, 558 and 520 on the number line.

(ii) Write the number each letter represents.

A = [ ]   B = [ ]   C = [ ]   D = [ ]   E = [ ]

(c) 247 is a [ ] -digit number.

The place with the greatest value is the _____ place.

The number is made up of [ ] hundreds, [ ] tens and [ ] ones.

(d) The greatest 2-digit number is [ ] .

The number that is 100 more than it is [ ] .

(e) (i)   387 + [ ] = 400

(ii) 650 − [ ] = 600

(iii) [ ] + 258 = 1000

(f) 2 hundreds and 18 tens make [ ] .

(g) Arrange the numbers 45, 405, 380, 1000, 806 and 968 in order, starting with the greatest:

[ ] [ ] [ ] [ ] [ ] [ ]

6 True or false? (Put a ✓ for true and a ✗ for false in each box.)

(a) The place with the greatest value in 65 is in the tens place, therefore it is a 10-digit number. . . . . ☐

(b) 5 hundreds and 4 ones make 504. . . . . ☐

(c) There are 37 tens in 370. . . . . ☐

(d) To make ◯ × 4 < 31 true, the ◯ can be filled in with any of the seven numbers 1, 2, 3, 4, 5, 6 and 7. ☐

7 Multiple choice questions. (For each question, choose the correct answer and write the letter in the box.)

(a) The place with the greatest value in a 3-digit number is ☐ .

A. ones place   B. tens place
C. hundreds place   D. thousands place

(b) Using ■ to represent hundreds and ● to represent ones,

■■■■■●●●●●● represents the number ☐ .

A. 46   B. 460   C. 604   D. 406

(c) The tens numbers that come before and after 786 are ☐ .

A. 785 and 787   B. 780 and 790
C. 700 and 800   D. 770 and 790

(d) One more dot is added into one of the columns of the place value chart below. The number represented **cannot** be ☐ .

Hundreds	Tens	Ones
●●	●●●	

A. 330   B. 240   C. 231   D. 241

8 A class in Year 3 has 30 pupils. For a school reading programme, the class is divided into 6 groups. Each group receives 9 books.

How many books does the class receive in total?

Number sentence: _____

Answer: _____

9 How many legs do 8 chicks and 3 rabbits have in total?

Number sentence: _____

Answer: _____

10 A fairground dodgem car can seat 4 children.
How many cars are needed for 30 children?

Number sentence: _____

Answer: _____

11 A sports club has 34 footballs, 18 fewer than the volleyballs. There are 28 more basketballs than volleyballs.

How many volleyballs does the club have?

How many basketballs are there?

Number sentence: _____

Answer: _____

Chapter 4 Statistics (II)

4.1 From statistical table to bar chart

Learning objective Interpret and represent data using tables and bar charts

Basic questions

1 60 children are divided into 3 groups for 3 different tasks.

Group A has 20 children, Group B has 30 children and Group C has 10 children. Present this information by completing the following:

(a) A statistical table

	Group A	Group B	Group C
Number of children			

(b) A pictogram

Group A	
Group B	
Group C	

Each ○ stands for 10 children.

(c) A block diagram

Each cell stands for 5 children.

35
30
25
20
15
10
5
0
Group A   Group B   Group C

2 Read the bar chart and complete the statistical table.

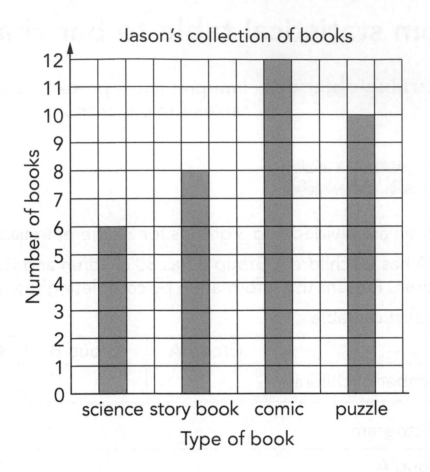

Jason's collection of books

Type of book	science	story book	comic	puzzle
Number of books				

3 Use the bar chart to complete the sentences. Each pupil chose only one favourite sport.

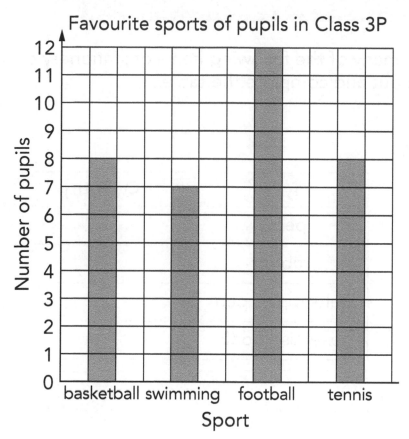

(a) In the graph, for the number of pupils, 1 unit represents ☐ pupil(s).

(b) _____ is chosen by the greatest number

of pupils and _____ is chosen by
the fewest pupils.

The difference is ☐ pupils.

(c) Two sports, _____

and _____, were chosen by the same
number of pupils.

(d) There are ☐ pupils in total in Class 3P.

Challenge and extension question

4 (a) How many of the following items of stationery do you have? Find out and complete the table.

Type	Quantity
pencil	
rubber	
felt-tipped pen	
exercise book	

(b) Now use another suitable statistical tool you have learned to present the data.

4.2 Bar charts (1)

Learning objective Interpret and represent data using bar charts

Basic questions

1 Read the bar chart and answer the questions.

Favourite fruit

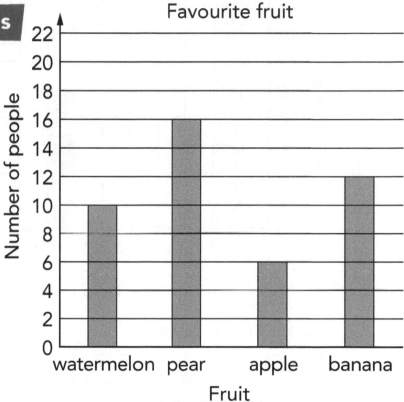

(a) In the graph, for the number of people, 1 unit represents ☐ people.

(b) Pear is ☐ people's favourite fruit while banana is ☐ people's favourite fruit.

(c) There are ☐ more people choosing watermelon than apple as their favourite fruit.

(d) There are ☐ people in total choosing watermelon and apple as their favourite fruit.

(e) There are ☐ times as many people choosing banana as their favourite fruit as those choosing apple.

2 Use the bar chart to complete the information. Each pupil in Class 3T chose only one favourite toy.

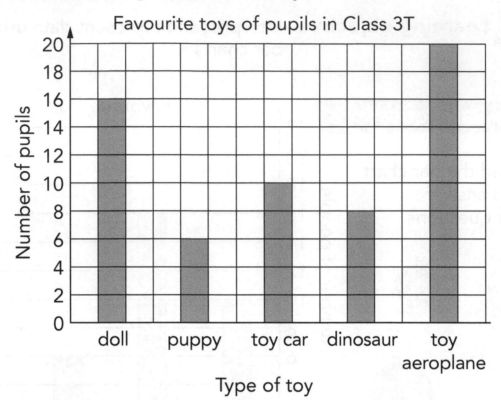

Favourite toys of pupils in Class 3T

(a) Complete the table.

Type	doll	puppy	toy car	dinosaur	toy aeroplane
Number of pupils					

(b) The toy favoured by the greatest number of pupils

is _____, and that by the fewest pupils

is _____.

The difference is ⬚ pupils.

(c) Based on the data above, there are ⬚ pupils in total in Class 3T.

3 Jenna did a class survey on the number of pupils taking part in PE activities and presented the data in the table below.

Complete a bar chart based on the data.

Type	rope skipping	dodgeball	football	running
Number of pupils	6	11	8	13

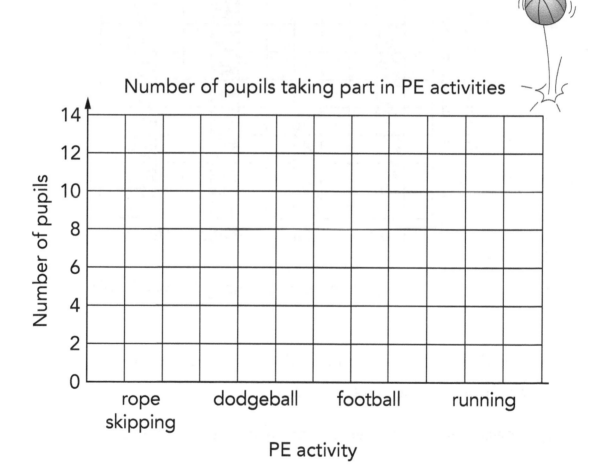

Number of pupils taking part in PE activities

Challenge and extension question

4 Read the bar chart and answer the questions. (Note: there are 4 pieces of fruit in each bag.)

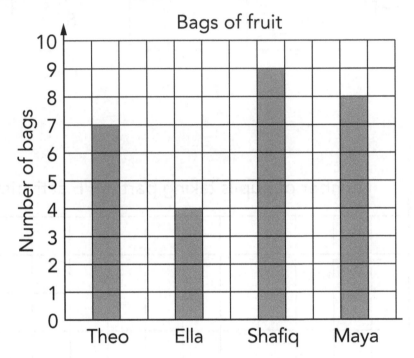

(a) _____ has twice as many bags of fruit

as _____.

(b) For the 4 children to have the same number of bags of fruit, each one must have ☐ bags.

(c) Theo has ☐ fewer pieces of fruit than Shafiq.

4.3 Bar charts (2)

Learning objective Interpret and represent data using tables, pictograms and bar charts

Basic questions

1 The pictogram shows information about the ages of the pupils who took part in a school activity.

Age	Number of pupils
11	🧍🧍🧍
10	🧍🧍
9	🧍
8	🧍 (half)
7	🧍🧍🧍
6	🧍🧍

Each 🧍 stands for 2 pupils.

(a) (i) How many pupils are aged 9? ☐

   (ii) How many are younger than 9? ☐

   (iii) How many are older than 9? ☐

(b) How many pupils in total took part in the school activity? ☐

99

(c) Complete a bar chart based on the information in the pictogram on page 99.

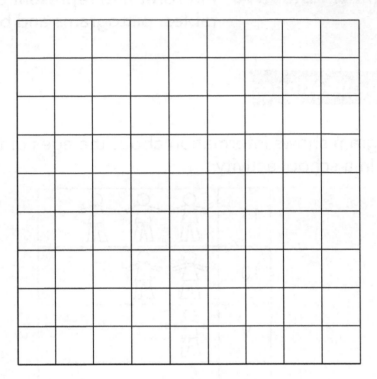

2 During PE, Asif, Tom, Joe, Sarah and Lila had a competition to see who could bounce a ball the most times in one minute. Their results are recorded in the table below.

Results of ball bouncing

Name	Asif	Tom	Joe	Sarah	Lila
Number of times	40	25	50	45	35

(a) Look at the table and put the children's results in order. Write the names, starting with the child with the highest score.

_____ _____ _____ _____ _____

(b) Show the results in a bar chart.

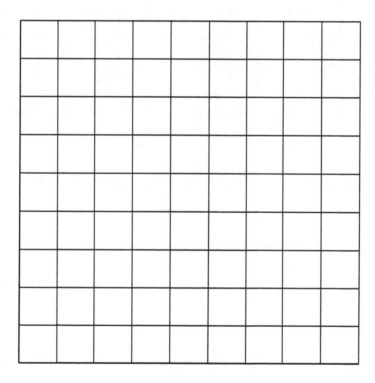

Challenge and extension question

3 Use the bar chart to complete the sentences. Each pupil takes part in only one activity.

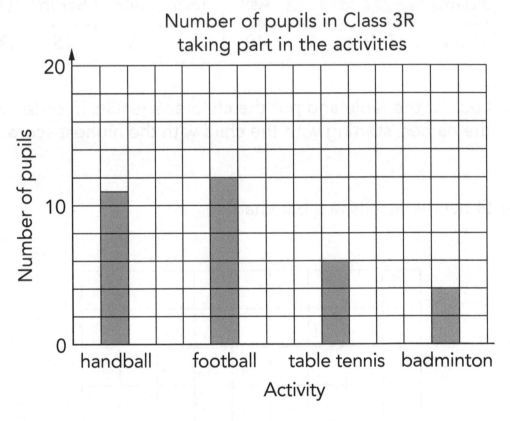

Number of pupils in Class 3R
taking part in the activities

(a) In the graph, for the number of pupils, 1 unit represents ☐ pupil(s).

(b) The activity with the greatest number of participants

   is _____ and that with the fewest participants

   is _____ .

(c) There are ☐ pupils taking part in all the activities in Class 3R.

(d) The difference between the number of pupils taking part in handball and the number taking part in football is ☐ .

Chapter 4 test

1 In the bar charts below, what does one cell stand for? What does each bar stand for?

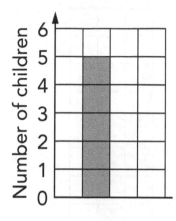

(a) One cell stands for _____ .

The bar stands for _____ .

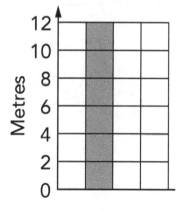

(b) One cell stands for _____ .

The bar stands for _____ .

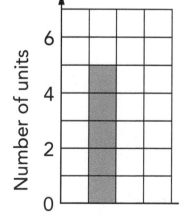

(c) One cell stands for _____ .

The bar stands for _____ .

2 The bar chart shows how Class 3S get to school.
Use the bar chart to answer the questions.

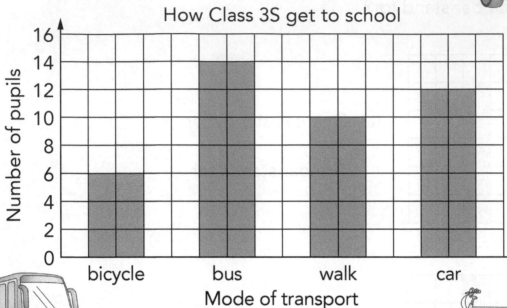

(a) In the graph, one cell stands for ☐ pupil(s).

(b) The greatest number of pupils go to school by _____.

The number is ☐.

The smallest number of pupils go to school by _____.

The number is ☐.

(c) How many pupils go to school by bus or car? ☐

(d) How many pupils are there in Class 3S altogether? ☐

3 Look at the bar chart.

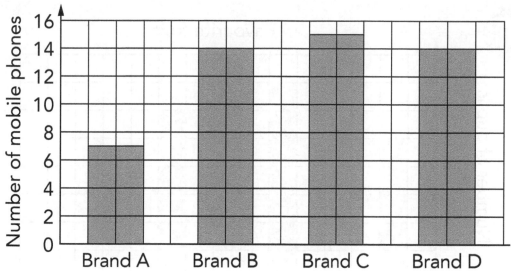

Sales of mobile phones in an online store in a week

(a) Use the bar chart to fill in the table.

Brand	Brand A	Brand B	Brand C	Brand D
Number of mobile phones sold				

(b) Based on the sales recorded, the most popular brand

is _____ and the least popular is _____ .

(c) Two brands, _____ and _____ , are
equally popular according to the sales in the week.

(d) The store sold ☐ mobile phones of the four brands in the week.

4 The bar chart below shows the favourite fruit of the pupils in a school.

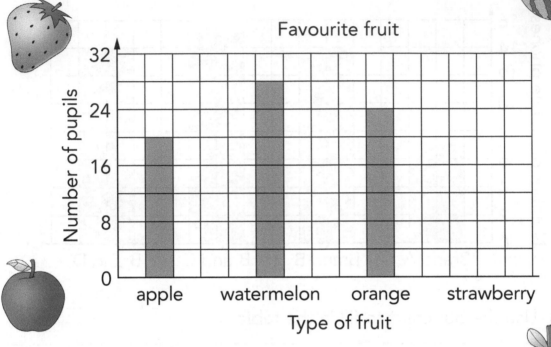

(a) In the graph, 1 unit of the number of pupils represents ☐ pupil(s).

There are ☐ pupils whose favourite fruit is watermelon.

☐ fewer pupils favour apple than favour orange.

(b) The favourite fruit of 14 pupils is strawberry.

Complete the bar chart to show this information.

(c) Write your own question based on the bar chart and give the answer.

_____

_____

_____

5.1 Second and minute

 Learning objective Understand second and minute as units of time

 Basic questions

1 Complete the sentences.

(a) There are ☐ seconds in a minute.

(b) There are ☐ minutes in an hour.

(c) There are ☐ seconds in half a minute.

(d) There are ☐ seconds in 10 minutes.

2 Draw lines to match the times.

| 03:05 | 08:26 | 04:59 | 11:11 |

3 Write a suitable unit of time in each space.

(a) It took Tom 48 _____ to swim 50 metres.

(b) Anya practises the piano 1 _____ every day.

(c) Ahmed's father works 8 _____ a day.

(d) The pupils have a 50 _____ lunch break.

4 Try the following activities and record the results.

(a) I can read ☐ English words in 1 minute.

(b) I can skip with a rope ☐ times in 1 minute.

(c) It takes me ☐ seconds to walk 50 metres.

(d) My heart beats ☐ times in 10 seconds.

5 Convert the times into different units.

(a) $\frac{1}{2}$ hour = ☐ minutes

(b) 5 minutes = ☐ seconds

(c) $\frac{3}{4}$ hour = ☐ minutes

(d) 120 seconds = ☐ minutes

(e) 1 hour 40 minutes = ☐ minutes

(f) 2 minutes 30 seconds = ☐ seconds

(g) 90 seconds = ☐ minute ☐ seconds

(h) 100 minutes = ☐ hour ☐ minutes

Challenge and extension questions

6 Complete the sentences with the correct times.
You may use a clock face to help you.

(a) It is now 2:30. After 1 hour, it will be [   ].

(b) It is now 5:35. After 2 hours, it will be [   ].

(c) It is now 11:28. After 5 minutes, it will be [   ].

(d) It is now 6:17. After 1 hour and 30 minutes, it will be [   ].

7 To the nearest minute, the time shown
on the clock face on the right is [   ].

Think of the position of the second hand and give a reason for
your answer.

_____

_____

5.2 Times on 12-hour and 24-hour clocks and in Roman numerals

Learning objective Read times in Roman numerals and convert between 12-hour and 24-hour times

Basic questions

1 Fill in the boxes.

(a) There are ☐ hours in a day.

(b) There are ☐ hours in half a day.

(c) The time at noon is ☐ o'clock.

(d) In a day, the time from midnight to noon is

called _____ or a.m.

(e) The time from noon to midnight is called _____ or p.m.

2 What number does each Roman numeral represent? The first one has been done for you.

IX = 9   I = ☐   X = ☐   II = ☐   VI = ☐   XI = ☐

XII = ☐   VII = ☐   III = ☐   V = ☐   IV = ☐   VIII = ☐

3 Read each clock and write the time in digits underneath.

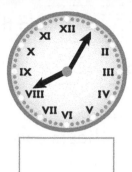

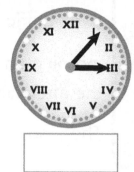

4 Complete the table for converting 12-hour time to 24-hour time. The first one has been done for you.

12-hour time	24-hour time
3:03 a.m.	03:03
8:00 a.m.	
	13:36
	23:58
12:00 midnight	

5 Read each clock and then write the 24-hour times in the boxes below.

(a)

(b)

(c)

or

or

or

6  The traffic road sign shows that the road is closed

to all traffic from ☐ to ☐ in the

morning and from ☐ to ☐ in

the afternoon.

07:30 – 10:30
16:30 – 19:30

The road is closed to traffic for ☐ hours each day.

 ## Challenge and extension question

7 Draw a line to match each pair of times that would look the same on the clock face. One has been done for you.

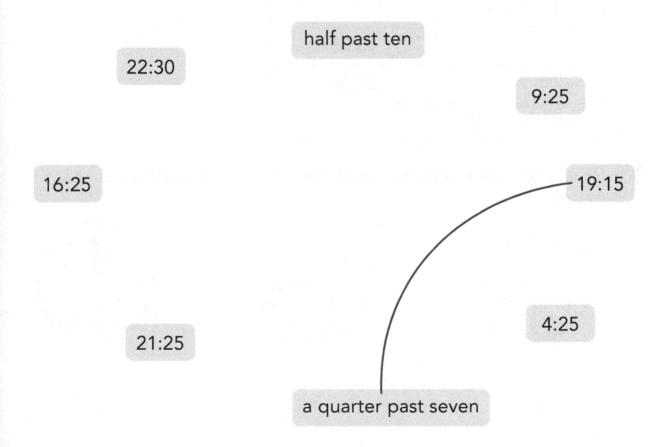

half past ten

22:30

9:25

16:25

19:15

21:25

4:25

a quarter past seven

5.3 Leap years and common years

Learning objective Know the number of days in each leap year and common year

Basic questions

1 True or false? (Put a ✓ for true and a ✗ for false in each box.)

(a) One day always has 24 hours. . . . . □

(b) One week always has 7 days. . . . . □

(c) One month always has 30 days. . . . . □

(d) One year always has 12 months. . . . . □

(e) One year always has 365 days. . . . . □

2 Use a calendar for this year to complete the table.

Month	Jan	Feb	Mar	Apr	May	Jun
Number of days						
Month	Jul	Aug	Sept	Oct	Nov	Dec
Number of days						

3 Use the information in the table in Question 2 to complete the facts.

(a) There are ☐ days in April. The months that have the same number of days as April are _____

_____ .

(b) Christmas Day is on the ☐ of _____ .

The months that have the same number of days as that month

are _____

_____ .

(c) The month that has the fewest days is _____ .

There are [ ] days in that month.

(d) There are [ ] days in this year.

4 Using a print or online calendar, find the number of days in each year and the number of days in February for any 10 consecutive years.

Year	Number of days in the year	Number of days in February

5 Use the information in the table in Question 4 to fill in the answers.

(a) There are ☐ days in most of the months of February. These years are called common years.

(b) There are ☐ days in some of the months of February. These years are called leap years.

(c) A common year has ☐ days. A leap year has ☐ days.

(d) 2016 was a leap year. The next two leap years will be ☐ and ☐ .

Challenge and extension question

6 Alvin is at primary school. He said: "I have only had three birthdays, including the day I was born."

On what day of the year do you think he was born?

5.4 Calculating the duration of time

Learning objective Compare the duration of events

Basic questions

1 Write >, < or = in each 〇.

(a) 1 hour 〇 60 seconds

(b) 2 minutes 〇 100 seconds

(c) 2 days 〇 20 hours

(d) 3 hours 〇 200 minutes

(e) 1 month 〇 27 days

(f) 52 weeks 〇 1 year

2 Find the answers.

(a) May left home at 9:00 a.m. and came back at 1:00 p.m.

She was away for ☐ hours.

(b) Lottie began her homework at 6:00 p.m. and finished at 6:45 p.m.

She spent ☐ minutes doing her homework.

(c) A party started at 19:30 and ended at 21:40.

It lasted ☐ hours and ☐ minutes.

(d) A volleyball match started at 19:30 and lasted 155 minutes.

It ended at ☐ .

3 ABC Superstore's opening hours on weekdays are shown below.

(a) The superstore opens at ⬚ in the morning and closes

   at ⬚ in the evening.

(b) The superstore is open for ⬚ hours and ⬚ minutes on
    weekdays.

(c) It takes Mum 20 minutes to go from work to the superstore.
   If she leaves work for the superstore at 8 p.m., does she

   still have time to do shopping? _____.

   If so, for how long? _____.

4 The 2012 Summer Olympics was held in London from 27 July 2012
to 12 August 2012.

What was the duration of the event? _____

5 A local computer store had three promotional sales in the first half of 2015. The dates are given below.

First promotion	23/02/2015 until 02/03/2015
Second promotion	23/03/2015 until 01/04/2015
Third promotion	25/04/2015 until 03/05/2015

(a) Which promotion lasted for the longest period of time?

_____

How long was it? _____

(b) Which promotion lasted for the shortest period of time?

_____

How long was it? _____

(c) For how many days did the store have a promotion in the first

half of 2015? _____

Challenge and extension question

6 A popular TV series was broadcast from Thursday 7 March 2013 to 4 April 2013. Two episodes of the series were broadcast every day from Monday to Thursday and one episode was broadcast on Saturdays and Sundays. It was not shown on Fridays.

The series had [ ] episodes and was broadcast for [ ] days.

1 Use digits and a colon to write the time shown on each clock face to the nearest minute.

(a)

(b)

(c)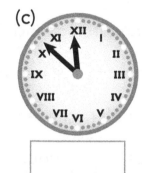

(d)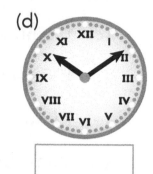

2 Draw the minute and hour hands on each clock face to show the time given.

(a)

(b)

(c)

(d)

3 Fill in the boxes.

(a) 2 minutes = ☐ seconds

(b) 2 days = ☐ hours

(c) 1 minute and 10 seconds = ☐ seconds

(d) 100 seconds = ☐ minute and ☐ seconds

(e) 1 hour and 30 minutes = ☐ minutes

(f) 85 minutes = ☐ hour and ☐ minutes

4 Write >, < or = in each ◯.

(a) 60 seconds ◯ 1 minute

(b) 1 minute and 40 seconds ◯ 100 seconds

(c) 100 minutes ◯ 1 hour

(d) 1 and a half days ◯ 30 hours

(e) 20 hours ◯ 1 days

(f) 1 and a quarter hours ◯ 90 minutes

5 True or false? (Put a ✓ for true and a ✗ for false in each box.)

(a) There are always 183 days in the first half of a year. . . . . . . ☐

(b) Compared with all the other months, February has the fewest days. . . . . ☐

(c) Uncle Joe works 35 hours every week. He works 140 hours in month. . . . . ☐

(d) If a month has 31 days, then the following month must have 30 days. . . . . ☐

6 Fill in the boxes.

(a) Aliyah left school at 3:13 p.m. After 25 minutes, she arrived home.

She reached home at ☐ .

(b) The duration of a flight from City A to City B was 2 hours. It arrived in City B at 1:40 p.m.

The flight took off from City A at ☐ .

(c) The football match ended at 20:00. It lasted 90 minutes.

The football match started at ☐ .

(d) A toy shop sold 7 toy cars each day. From 30 August to 10 September, it sold ☐ toy cars in total.

7 Mo and his family went on holiday. They were away for 2 days fewer than 2 weeks. How many days did they spend on holiday?

_____

8 Write the time shown on each clock face in the boxes above the clocks. Then fill in the boxes below the clocks with the duration from the time shown on one clock face to the time shown on the next.

(a) ☐   (b) ☐   (c) ☐   (d) ☐

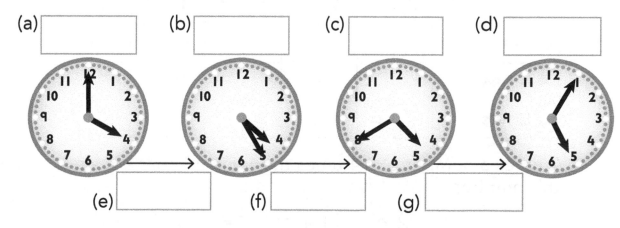

(e) ☐   (f) ☐   (g) ☐

Chapter 6 Consolidation and enhancement

6.1 5 threes plus 3 threes equals 8 threes

 Learning objective Solve problems involving multiplication and addition

 Basic questions

1 Look at the pictures and complete the number sentences.

(a) How many apples and pears are there altogether?

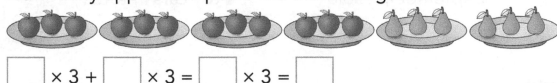

$\boxed{} \times 3 + \boxed{} \times 3 = \boxed{} \times 3 = \boxed{}$

(b) How many pineapples and apples are there altogether?

$\boxed{} \times \boxed{} + \boxed{} \times \boxed{} = \boxed{} \times \boxed{} = \boxed{}$

2 Complete the table and then fill in the answers below.

	1	2	3	4	5	6	7	8	9	10
2 times										
4 times										
6 times										

2 times a number plus 4 times the same number equals $\boxed{}$ times this number.

$$2 \times \boxed{} + 4 \times \boxed{} = \boxed{} \times \boxed{}$$

3 Fill in the boxes.

(a) $3 \times 6 + 5 \times 6 = \boxed{} \times 6 = \boxed{}$

(b) $3 \times 4 + 6 \times 4 = \boxed{} \times 4 = \boxed{}$

(c) $8 \times 2 + 2 \times 2 = 10 \times \boxed{} = \boxed{}$

(d) $7 \times 7 + 2 \times 7 = 7 \times \boxed{} = \boxed{}$

4 Calculate mentally.

(a) $4 \times 8 + 2 \times 8 = \boxed{}$

(b) $5 \times 7 + 4 \times 7 = \boxed{}$

(c) $5 \times 9 + 5 \times 9 = \boxed{}$

(d) $6 \times 2 + 3 \times 2 = \boxed{}$

5 Think carefully. What number can you write in each box?

(a) $8 \times 3 + \boxed{} \times 3 = 9 \times 3$

(b) $4 \times 6 + \boxed{} \times 6 = 6 \times 6$

(c) $\boxed{} \times 5 + 2 \times 5 = 5 \times 5$

(d) $2 \times \boxed{} + 6 \times \boxed{} = 8 \times 7$

6 Each pen costs £9. Tom bought 2 pens and Joana bought 3 pens. How much did they pay in total?

£9

Challenge and extension questions

7 Think carefully and then fill in the boxes.

(a) $4 \times 8 + 8 = \boxed{} \times 8$

(b) $6 \times 5 + \boxed{} = 7 \times 5$

8 Show different ways to express 9×5.

(a) $9 \times 5 = \boxed{} \times \boxed{} + \boxed{} \times \boxed{}$

(b) $9 \times 5 = \boxed{} \times \boxed{} + \boxed{} \times \boxed{}$

(c) $9 \times 5 = \boxed{} \times \boxed{} + \boxed{} \times \boxed{}$

(d) $9 \times 5 = \boxed{} \times \boxed{} + \boxed{} \times \boxed{}$

(e) $9 \times 5 = \boxed{} \times \boxed{} + \boxed{} \times \boxed{} + \boxed{} \times \boxed{}$

6.2 5 threes minus 3 threes equals 2 threes

Learning objective Solve problems involving multiplication and subtraction

Basic questions

1 Look at the pictures, write number sentences and then calculate.

(a) How many more pineapples are there than apples?

☐ × 2 − ☐ × 2 = ☐ × 2 = ☐

(b) How many are left over?

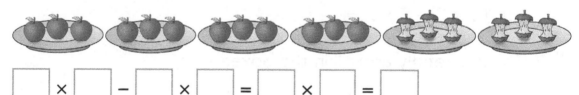

☐ × ☐ − ☐ × ☐ = ☐ × ☐ = ☐

2 Complete the table and then fill in the answers below.

	1	2	3	4	5	6	7	8	9	10
9 times										
5 times										
4 times										

9 times a number minus 5 times the same number equals ☐ times this number.

$$9 \times \boxed{} - 5 \times \boxed{} = \boxed{} \times \boxed{}$$

Consolidation and enhancement

3 Fill in the boxes.

(a) $8 \times 6 - 5 \times 6 = \boxed{} \times 6 = \boxed{}$

(b) $7 \times 4 - 6 \times 4 = 4 \times \boxed{} = \boxed{}$

(c) $5 \times 9 - 5 \times 6 = 5 \times \boxed{} = \boxed{}$

(d) $5 \times 7 - 2 \times 7 = \boxed{} \times \boxed{} = \boxed{}$

(e) $9 \times 6 - 4 \times \boxed{} = \boxed{} \times 6 = \boxed{}$

4 Calculate mentally.

(a) $12 \times 8 - 10 \times 8 = \boxed{}$

(b) $15 \times 7 - 9 \times 7 = \boxed{}$

(c) $18 \times 9 - 9 \times 9 = \boxed{}$

(d) $16 \times 2 - 7 \times 2 = \boxed{}$

5 Think carefully and fill in the boxes.

(a) $8 \times 3 - \boxed{} \times 3 = 6 \times 3$

(b) $12 \times 6 - \boxed{} \times 6 = 9 \times 6$

(c) $\boxed{} \times 5 - 6 \times 5 = 4 \times 5$

(d) $10 \times \boxed{} - 6 \times \boxed{} = \boxed{} \times 7$

6 Each notebook costs £9. Tom bought 8 books and Joana bought 3 books. How much more did Tom pay than Joana?

Challenge and extension questions

7 Think carefully and then fill in the boxes.

(a) $8 \times 8 - 8 = \boxed{} \times \boxed{}$

(b) $8 \times 5 = 12 \times 5 - \boxed{} \times \boxed{}$

(c) $8 \times 7 - 2 \times 7 - 4 \times 7 = 7 \times \boxed{}$

(d) $2 \times 2 = 9 \times 2 - 4 \times \boxed{} - \boxed{} \times \boxed{}$

8 Use different ways to express 3×6.

(a) $3 \times 6 = \boxed{} \times \boxed{} - \boxed{} \times \boxed{}$

(b) $3 \times 6 = \boxed{} \times \boxed{} - \boxed{} \times \boxed{}$

(c) $3 \times 6 = \boxed{} \times \boxed{} - \boxed{} \times \boxed{}$

(d) $3 \times 6 = \boxed{} \times \boxed{} - \boxed{} \times \boxed{} - \boxed{} \times \boxed{}$

6.3 Multiplication and division

Learning objective Use the relationship between multiplication and division to solve problems

Basic questions

1 Calculate mentally.

(a) $4 \times 6 = \boxed{}$

(b) $7 \times 8 = \boxed{}$

(c) $5 \times 9 = \boxed{}$

(d) $10 \times 10 = \boxed{}$

(e) $24 \div 4 = \boxed{}$

(f) $56 \div 7 = \boxed{}$

(g) $45 \div 9 = \boxed{}$

(h) $100 \div 10 = \boxed{}$

(i) $24 \div 6 = \boxed{}$

(j) $56 \div 8 = \boxed{}$

(k) $45 \div 5 = \boxed{}$

(l) $0 \times 7 = \boxed{}$

2 What is the greatest number you can write in each box?

(a) $\boxed{} \times 5 < 42$

(b) $\boxed{} \times 6 < 37$

(c) $\boxed{} \times 8 < 80$

(d) $7 \times \boxed{} < 56$

(e) $48 > 9 \times \boxed{}$

(f) $58 > 6 \times \boxed{}$

3 Draw lines to match the conditions to the questions. Then write the number sentences and answer the questions.

Each set consists of one desk and one chair.

There are 6 rows of desks and chairs in the classroom. There are 7 sets in each row.

How many rows are there?

There are 42 sets of desks and chairs. There are 7 sets in each row.

How many sets are there in each row?

There are 42 sets of desks and chairs. They are put into 6 rows equally.

How many sets of desks and chairs are there?

(a) Number sentence: _____

There are ☐ rows.

(b) Number sentence: _____

There are ☐ sets in each row.

(c) Number sentence: _____

There are ☐ sets of desks and chairs.

4 Application problems.

(a) There are 6 sheep on the hillside.
There are 6 times as many deer as sheep.
How many deer are there? ☐

(b) 10 boys made some paper models. Each of them made 4 paper models. A group of girls made 32 paper models in total. Who made more, the boys or the girls?

_____

(c) A taxi can seat 4 passengers.
There are 27 passengers.
What is the smallest number of taxis needed to carry all the passengers? ☐

(d) The month of July has ☐ weeks and ☐ days.

The division sentence that shows this is:

_____ .

Challenge and extension questions

5 There are fewer than 30 cupcakes in a box. If they are divided equally among 4 or 5 children, there will always be 1 cupcake left over.

How many cupcakes are there? ☐

6 The cost of 1 = the cost of 8 .

The cost of 4 pairs of = the cost of 8 .

The cost of 2 = the cost of ☐ .

6.4 Mathematics plaza – dots and patterns

Learning objective Explore patterns of odd and even numbers

Basic questions

1 Look at the dot diagrams and write 'even' or 'odd' in each box.

(Remember: even numbers end in 0, 2, 4, 6 and 8, and odd numbers end in 1, 3, 5, 7 and 9.)

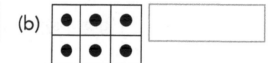

(a) ⬚⬚⬚⬚ / ⬚⬚⬚⬚ ☐

(b) ⬚⬚⬚ / ⬚⬚⬚ ☐

(c) ⬚⬚⬚⬚ / ⬚⬚⬚ ☐

(d) ⬚⬚ / ⬚⬚⬚ ☐

2 Look at the dot diagrams and write the number sentences.

(a) ☐ + ☐ = ☐

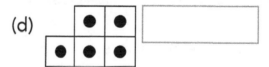

(b) ☐ + ☐ = ☐

(c) ☐ + ☐ = ☐

3 Find the pattern and then fill in the boxes.

(a) 1, 3, 5, 7, 9, ☐ , ☐ , ☐

(b) 2, 4, 6, 8, 10, ☐ , ☐ , ☐

(c) 30, 28, 26, 24, 22, ☐ , ☐ , ☐

(d) 2, 5, 4, 7, 6, 9, 8, 11, ☐ , ☐ , ☐ , ☐

4 Look at the diagram and then fill in the boxes.

(a) 16 ÷ 3 = ☐ (groups) with a remainder of ☐ (circle)

(b) 16 ÷ 5 = ☐ (circles) with a remainder of ☐ (circle)

5 Write the numbers.

(a) 5 odd numbers: ☐ ☐ ☐ ☐ ☐

(b) 5 even numbers: ☐ ☐ ☐ ☐ ☐

(c) All 2-digit odd numbers with a 3 in the tens place:

_____

(d) All 2-digit even numbers with a 6 in the tens place:

_____

(e) Four 2-digit even numbers after 19: ☐ ☐ ☐ ☐

6 Think carefully and fill in the boxes.

(a) $1 + 3 = 2 \times 2 = \boxed{}$

(b) $1 + 3 + 5 = 3 \times 3 = \boxed{}$

(c) $1 + 3 + 5 + 7 = 4 \times \boxed{} = \boxed{}$

(d) $1 + 3 + 5 + 7 + 9 = \boxed{} \times \boxed{} = \boxed{}$

(e) $1 + 3 + 5 + 7 + \boxed{} + \boxed{} = \boxed{} \times \boxed{} = \boxed{}$

Challenge and extension questions

7 Complete the calculations.

(a) $8 + 1 = \boxed{}$ (b) $8 + 2 = \boxed{}$ (c) $9 + 1 = \boxed{}$

(d) $8 + 3 = \boxed{}$ (e) $8 + 4 = \boxed{}$ (f) $9 + 3 = \boxed{}$

(g) $8 + 5 = \boxed{}$ (h) $8 + 6 = \boxed{}$ (i) $9 + 5 = \boxed{}$

(j) $8 + 7 = \boxed{}$ (k) $8 + 8 = \boxed{}$ (l) $9 + 7 = \boxed{}$

8 Think carefully. Then write 'odd number' or 'even number' in each answer space.

(a) even number + odd number = _____

(b) even number + even number = _____

(c) odd number + odd number = _____

6.5 Mathematics plaza – magic square*

Learning objective Explore patterns on magic squares

Basic questions

1 Work out the answer for each box.
Two have been done for you.

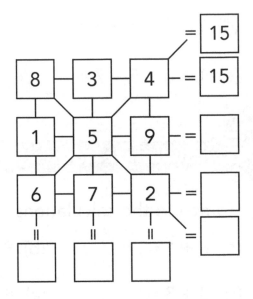

2 Which of these is a magic square? Put a ✓ for yes and a ✗ for no in the small box underneath.

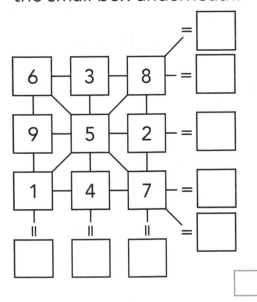
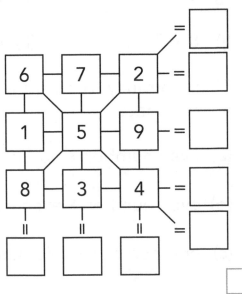

* In a magic square, each number is used once and all the numbers in every row, column and diagonal add up to the same number.

3 Fill in each box with a number so that the sum of the three numbers on each line is 15.

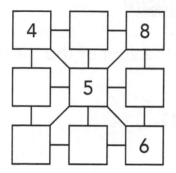

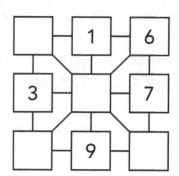

4 Fill in the cells with different numbers so that all three numbers in each row, each column and each diagonal add up to the number in the circle above.

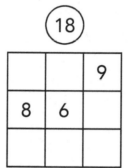

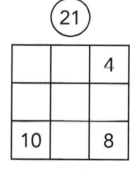

5 Fill in each cell with a suitable number so that the sum of the three numbers in each row and column is 10.

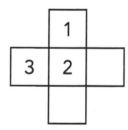

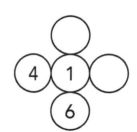

Challenge and extension question

6 Put these numbers in the circles so that the sum of the three
numbers on each arm is the same.

2   4   6   8   10   12   14

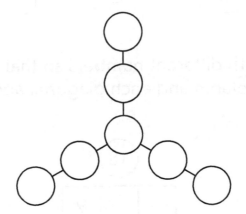

6.6 Numbers to 1000 and beyond

Learning objective Recognise and use place value of 3-digit and 4-digit numbers

Basic questions

1 Calculate mentally.

(a) 57 − 33 + 45 = ☐

(b) 1 × 6 × 6 = ☐

(c) 64 ÷ 8 − 0 = ☐

(d) 23 + 82 + 92 = ☐

(e) 6 × 6 × 0 = ☐

(f) 50 ÷ 5 + 72 = ☐

2 Complete the place value chart.

_____ place	_____ place	_____ place	*tens* place	*ones* place

3 Complete the number pattern.

1000, 2000, ☐, 4000, ☐, 6000, ☐, 8000, ☐, 10000, ☐, 12000

4 Write the digits of each number in the place value chart. The first one has been done for you.

(a) 4208

TTh	Th	H	T	O
0	4	2	0	8

(b) 9990

TTh	Th	H	T	O

(c) 10 008

TTh	Th	H	T	O

(d) 2006

TTh	Th	H	T	O

5 Write the numbers represented in each place value chart in words and numerals.

(a)

TTh	Th	H	T	O
	●●	●●		●●

In words: _____

In numerals: _____

(b)

TTh	Th	H	T	O
●				

In words: _____

In numerals: _____

(c)

TTh	Th	H	T	O
	●●●●●	●●	●●●	

In words: _____

In numerals: _____

(d)

TTh	Th	H	T	O
	●●●●		●●●●●	●●●

In words: _____

In numerals: _____

6 Complete each sentence.

(a) 10 ones make ☐, 10 tens make ☐, 10 hundreds make ☐ and 10 thousands make ☐.

(b) Counting from the right, the first digit of a number is its _____ place. The third digit is its _____ place. The fifth digit is its _____ place.

(c) A number consisting of 7 thousands, 5 hundreds, 2 tens and 3 ones is ☐.

(d) There are ☐ thousands or ☐ hundreds in 6000. There are ☐ tens in 170.

(e) Three consecutive numbers after 9998 are ☐, ☐, and ☐.

Challenge and extension question

7 Complete the number patterns.

(a) 5078, 5079, ☐, ☐, 5082

(b) 2323, 3434, 4545, ☐, ☐, 7878

(c) 10 000, 9990, 9980, ☐, ☐, ☐

6.7 Read, write and compare numbers to 1000 and beyond

Learning objective Read, write and compare numbers beyond 1000

Basic questions

1 Calculate mentally.

(a) 70 − 3 + 13 = ☐

(b) 53 − 37 − 4 = ☐

(c) 6 × 8 + 3 = ☐

(d) 47 − 11 − 21 = ☐

(e) 6 × 4 + 61 = ☐

(f) 40 ÷ 4 − 5 = ☐

(g) 90 ÷ 9 × 5 = ☐

(h) 23 + 28 + 92 = ☐

(i) 4 × 8 + 6 = ☐

(j) 56 ÷ 8 × 6 = ☐

2 Read and write the numbers in words and numerals to complete the table.

Words	Numerals
six thousand three hundred and forty-eight	
five thousand and fifty	
thirteen thousand and four	
	9008
	4415
	19 006

Consolidation and enhancement

3 Read and write the numbers in words and then fill in the boxes.

(a) 4632 _____

4632 = ☐ + ☐ + ☐ + ☐

(b) 2547 _____

2547 = ☐ + ☐ + ☐ + ☐

(c) 6003 _____

6003 = ☐ + ☐ + ☐ + ☐

(d) 2030 _____

2030 = ☐ + ☐ + ☐ + ☐

4 Write the numbers in numerals.

(a) One thousand eight hundred and twelve: ☐

(b) Four thousand and fifty: ☐

(c) Six thousand five hundred: ☐

(d) Five thousand and six: ☐

5 Write >, < or = in each ◯.

(a) 985 ◯ 895   (b) 1000 ◯ 999   (c) 7801 ◯ 7081

(d) 3877 ◯ 3787   (e) 5020 ◯ 2050   (f) 3456 ◯ 3546

(g) 5420 ◯ 5421   (h) 9887 ◯ 9987   (i) 4002 ◯ 4200

6 Multiple choice questions. (For each question, choose the correct answer and write the letter in the box.)

(a) Four thousand and five hundred is written as ☐.

    A. 450   **B.** 4500   **C.** 4050

(b) Three thousand and seventeen is written as ☐.

    A. 307   **B.** 3007   **C.** 3017

(c) Nine thousand and three is written as ☐.

    A. 9030   **B.** 903   **C.** 9003

7 Use < to put the numbers in order, from the smallest to the greatest.

(a)   367   209   627   736

  ☐ < ☐ < ☐ < ☐

(b)   8070 8007 8700 7800

  ☐ < ☐ < ☐ < ☐

Challenge and extension question

8   5   4   0   0   9

Use these five digits to write the numbers below.

(a) The greatest 5-digit number: ☐

(b) Three numbers with 0 in the ones place:

☐ ☐ ☐

(c) Three numbers that do not have 0 in the ones place:

☐ ☐ ☐

(d) Three numbers greater than 90 000:

☐ ☐ ☐

Chapter 6 test

1 Calculate mentally.

(a) $6 \times 4 = \boxed{}$

(b) $9 \times 5 = \boxed{}$

(c) $32 \div 7 = \boxed{}$

(d) $30 \div 5 \times 4 = \boxed{}$

(e) $9 \times 7 = \boxed{}$

(f) $66 \div 8 = \boxed{}$

(g) $10 \times 10 = \boxed{}$

(h) $3 \times 8 + 4 \times 8 = \boxed{}$

(i) $36 \div 6 = \boxed{}$

(j) $0 \div 9 = \boxed{}$

(k) $4 \times 9 - 11 = \boxed{}$

(l) $7 \times 9 - 4 \times 9 = \boxed{}$

(m) $\boxed{} \div 10 = 3 \text{ r } 6$

(n) $19 \div \boxed{} = 3 \text{ r } 1$

2 What is the greatest number you can write in each box?

(a) $7 \times \boxed{} < 45$

(b) $31 > 6 \times \boxed{}$

(c) $\boxed{} \times 8 < 40$

(d) $20 > \boxed{} \times 9$

3 Write +, −, × or ÷ in each $\bigcirc$.

(a) $5 \bigcirc 6 = 11$

(b) $30 \bigcirc 0 = 0$

(c) $9 \bigcirc 3 = 3$

(d) $10 \bigcirc 3 = 30$

(e) $18 \bigcirc 18 = 0$

(f) $30 \bigcirc 5 = 6$

4 Write >, < or = in each $\bigcirc$.

(a) $2 \times 9 \bigcirc 3 \times 6$

(b) $9 \times 9 \bigcirc 9 + 9$

(c) $4 + 26 \bigcirc 66 - 30$

5 Fill in the boxes.

(a) 2 × 8 + 7 × 8 = ☐ × 8 = ☐

(b) 10 × 6 − 4 × 6 = ☐ × 6 = ☐

(c) 7 × ☐ < 8 × 6

(d) 42 ÷ 6 = ☐ ÷ 8

(e) 40 ÷ 5 > ☐ × 3

6 Complete the diagrams.

(a) Identify the patterns and fill in the diamonds with suitable multiplication sentences.

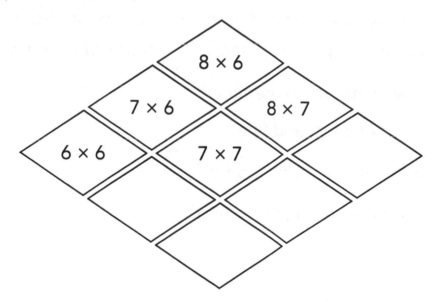

(b) Fill in each cell with a suitable number so that all three numbers in each row, column and diagonal total 15.

6		
	5	3

7 Fill in the answers.

(a) $1 + 3 + 5 + 7 + 9 =$ ☐ $\times 5 =$ ☐

(b) Given ▲ ÷ ● = 5 r 5, the smallest possible number that ● can be is ☐ . Then ▲ will be ☐ .

(c) If ● + ● + ▲ = 26 and ▲ × 5 = 40, then ● × ▲ = ☐ .

8 Solve the problems.

There are 8 cupcakes in each box.

(a) Some girls get 24 cupcakes, which is ☐ boxes.

(b) Some boys get 4 boxes, which is ☐ cupcakes.

(c) There are 22 girls and 18 boys in a class. If each pupil gets 1 cupcake, then they need ☐ boxes of cupcakes altogether.

9 Write the numbers in numerals.

(a) Four thousand and five ☐

(b) Six thousand eight hundred ☐

(c) One thousand and fifty-two ☐

(d) Ten thousand and thirty-nine ☐

10 Fill in the answers.

(a) 1005 consists of ☐ thousand and ☐ ones. It is written in words as: _____

(b) Put the following numbers in order.

1005, 105, 1050, 501, 5001.

☐ < ☐ < ☐ < ☐ < ☐

11 Use the digits 0, 0, 2 and 4 to write 4-digit numbers.

(a) The greatest 4-digit number: ☐

(b) The smallest 4-digit number: ☐

(c) Two different numbers with zero in the hundreds places:

☐ , ☐

(d) Four different numbers with zero in the ones places:

☐ , ☐ , ☐ , ☐

12 Application problems.

(a) 6 cups of water can fill up 1 kettle. 5 kettles of water can fill up 1 bucket. How many cups of water can fill up 1 bucket? ☐

(b) There are 8 footballs and 32 basketballs in a storage room. How many times as many basketballs are there as footballs? ☐

(c) Ms Smith and 26 pupils go rowing. If one boat can seat 6 people, how many boats do they need? ☐

(d) Shanaz wants to buy some yogurts with £17. Each pot of yogurt costs £2. Does she have enough money to buy 9 pots?

_____

(e) A group of boys got 6 plates of apples. Each plate had 5 apples. A group of girls got 32 apples. Who got more apples, the boys or the girls?

_____

13 The table shows the number of pupils in two Year 3 classes who are members of different sports clubs.

Sports club	badminton	softball	swimming	gymnastics
Number of pupils	6	12	18	12

(a) Show the results from the table in a bar chart. Remember to show the number of pupils that 1 unit in your bar chart stands for.

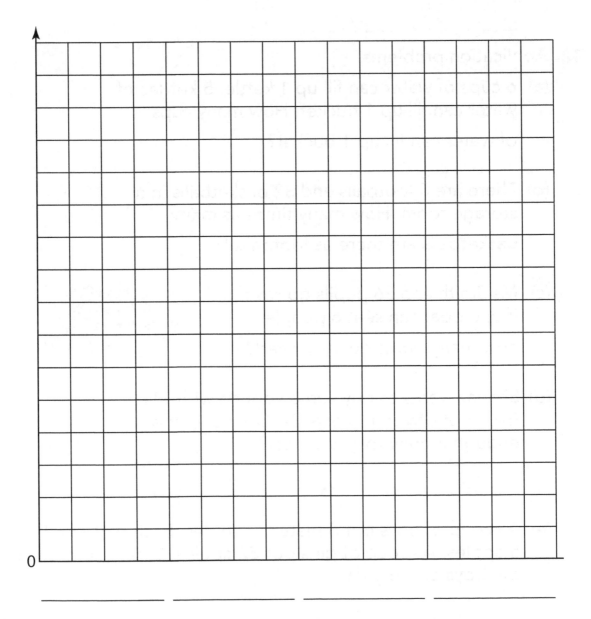

In the graph, 1 unit of the number of pupils stands for ☐ pupils.

(b) There are ☐ fewer Year 3 pupils in the badminton club than in the gymnastics club.

(c) Two clubs are equally popular in terms of members. They are the _____ club and the _____ club.

(d) The most popular club with Year 3 pupils is the _____ club. It has ☐ times as many members as the least popular club, which is the _____ club.

(e) There are ☐ pupils in total in the two classes who are members of the four sports clubs.

Notes

Notes

Notes

Notes

Notes

Notes

Notes